Data-Driven Optimization and Knowledge Discovery for an Enterprise Information System

Qing Duan • Krishnendu Chakrabarty • Jun Zeng

Data-Driven Optimization and Knowledge Discovery for an Enterprise Information System

 Springer

Qing Duan
PayPal, Inc.
San Jose, CA, USA

Jun Zeng
Hewlett-Packard Labs
Palo Alto, CA, USA

Krishnendu Chakrabarty
ECE
Duke University
Durham, NC, USA

ISBN 978-3-319-36429-2   ISBN 978-3-319-18738-9   (eBook)
DOI 10.1007/978-3-319-18738-9

Printed on acid-free paper

Springer International Publishing AG Switzerland is part of Springer Science+Business Media (www.springer.com)

To my parents for their unconditional love.
– Qing Duan

Preface

An enterprise information system (EIS) is an integrated data-applications platform characterized by diverse, heterogeneous, and distributed data sources. For many enterprises, a number of business processes still depend heavily on static rule-based methods and extensive human expertise. Enterprises are faced with the need for optimizing operation scheduling, improving resource utilization, discovering useful knowledge, and making data-driven decisions.

This book is focused on real-time optimization and knowledge discovery that addresses workflow optimization, resource allocation, as well as data-driven predictions of process-execution times, order fulfillment, and enterprise service-level performance. In contrast to prior work on data analytics techniques for enterprise performance optimization, the emphasis here is on realizing scalable and real-time enterprise intelligence based on a combination of heterogeneous system simulation, combinatorial optimization, machine-learning algorithms, and statistical methods.

On-demand digital print service is a representative enterprise requiring a powerful EIS. We use real data from a digital print service provider (PSP) to evaluate our optimization algorithms.

In order to handle the increase in volume and diversity of demands, we first present a high-performance, scalable, and real-time production scheduling algorithm for production automation based on an incremental genetic algorithm (IGA). The objective of this algorithm is to optimize the order dispatching sequence and balance resource utilization. Compared to prior work, this solution is scalable for a high volume of orders, and it provides fast scheduling solutions for orders that require complex fulfillment procedures. Experimental results highlight its potential benefit in reducing production inefficiencies and enhancing the productivity of an enterprise.

We next discuss analysis and prediction of different attributes involved in hierarchical components of an enterprise. We start from a study of the fundamental processes related to real-time prediction. Our process-execution time and process status prediction models integrate statistical methods with machine-learning algorithms. In addition to improved prediction accuracy compared to stand-alone machine-learning algorithms, it also performs a probabilistic estimation of the predicted status. An order generally consists of multiple series and parallel processes. We next introduce an order-fulfillment prediction model that combines advantages of multiple classification models by incorporating flexible decision-integration mechanisms. Experimental results show that adopting due dates recommended by the model can significantly reduce enterprise late-delivery ratio. Finally, we investigate service-level attributes that reflect the overall performance of an enterprise. We analyze and decompose time-series data into different components according to their hierarchical periodic nature, perform correlation analysis, and develop univariate prediction models for each component as well as multivariate models for correlated components. Predictions for the original time series are aggregated from the predictions of its components. In addition to a significant increase in mid-term prediction accuracy, this distributed modeling strategy also improves short-term time-series prediction accuracy.

In summary, the research described in this book has led to a set of characterization, optimization, and prediction tools for an EIS to derive insightful knowledge from data and use them as guidance for production management. It is expected to provide solutions for enterprises to increase reconfigurability, accomplish more automated procedures, and obtain data-driven recommendations or effective decisions.

The authors acknowledge all the support received from HP Labs. In particular, we would like to thank Production Automation Lab director Gary Dispoto for his continuing support for this research. We also would like to thank our research partners at RPI (http://www.rpiprint.com/, Seattle, WA) for providing the factory production dataset. Qing Duan acknowledges her advisor Professor Krishnendu Chakrabarty and mentor Dr. Jun Zeng who have brought her into this exciting data science field. Qing Duan also thanks all the support received from her labmates at Duke University, Fangming Ye, Kai Hu, Ran Wang, Zipeng Li, and Abhishek Koneru, for valuable discussions and collaboration.

San Jose, USA
Durham, USA
Palo Alto, USA

Qing Duan
Krishnendu Chakrabarty
Jun Zeng

Contents

Chapter 1
Introduction

An enterprise is a business entity that comprises of individuals that share a common purpose and utilize their skills and resources to achieve specific and declared goals. Enterprise goals, organizational structures, business processes, as well as information infrastructures define the business scope of an enterprise and its relationship with the external environment [1]. An overview diagram of the hierarchical structure of a production enterprise is shown in Fig. 1.1. At the highest level, a company or an affiliate is an entity that has to report its finances to the government on a timely basis. A plant can be a physical or a virtual place where goods are manufactured or traded. An enterprise may have its business presence across different locations. Its sales organization is responsible for selling goods as well as handling returns, credits, and materials at these locations. A distribution channel is the sales channel through which products are distributed, e.g., retail and wholesale. A division is purely related to the same type of product/service an enterprise is selling, e.g., hardware and software divisions. A storage location is the place where goods are stored before being sold. Each plant can be associated with multiple storage locations. It can be seen that in an enterprise, organizations or functional blocks are organized in a hierarchical manner [1].

In an enterprise, the fundamental activities are processes executed by resources. Parallel and series processes constitute different workflows. The completion of a service or an order requires the successful fulfillment of multiple workflows. As long as activities are occurring in an enterprise, various types of data are continuously generated and accumulated at different sites [3]. The resulting production and service environment of an enterprise is distributed, heterogeneous, and stochastic [4].

It is critical for an enterprise to acquire a better understanding of the commercial context of its organizations, such as customers, the market, supply and resources, and even competitors. The enterprise information system (EIS) is designed for such purposes. An EIS is a large-scale, integrated application-software layer that uses the computational, data storage, and data transmission power of modern information technology to support business processes, information flows, reporting,

© Springer International Publishing Switzerland 2015
Q. Duan et al., *Data-Driven Optimization and Knowledge Discovery
for an Enterprise Information System*, DOI 10.1007/978-3-319-18738-9_1

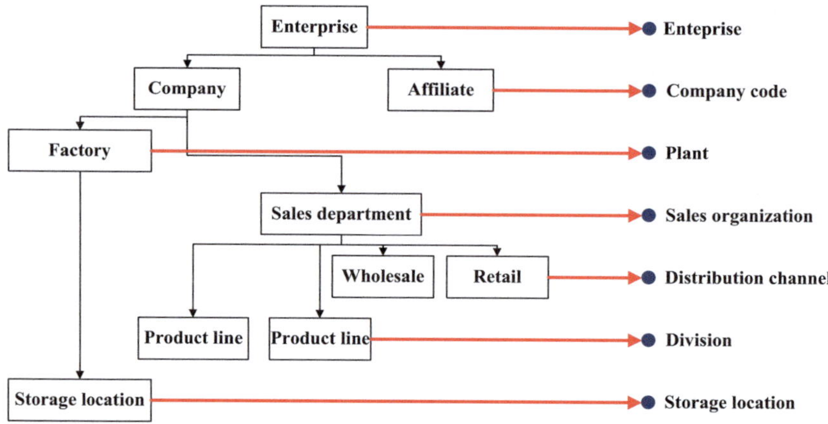

Fig. 1.1 An overview of the hierarchical structure of a production enterprise [2]

and data analytics within the enterprise [1, 5]. Although different enterprises have their own business processes and IT operations, the EIS they adopt fundamentally share the same design principles in building its architecture [3]. An intelligent EIS is responsible for providing historical, current, and predictive views of various business operations in order to provide a high quality of service to customers. For instance, attribute characterization, online learning, real-time decision making (e.g., dynamic scheduling and resource provisioning), performance management, workflow management, and predictive analytics. In summary, without the EIS performing operation optimization and knowledge discovery, an enterprise may not be able to improve production efficiency, retain valuable customers, or perform effective market analysis [5].

Data-driven techniques enable an EIS to be self-learning, as well as adaptive to dynamic service requests and resource availability. These techniques can be integrated into the EIS to cooperatively provide enterprise-wide optimization [6]. Our goal is to develop data-driven techniques for designing, managing, and transforming an enterprise to realize a higher level of agility, performance, and operational efficiency. The key ideas behind these techniques have originated from different disciplines and has the potential to significantly improve the performance of enterprises.

This research vision and research components are highlighted in Fig. 1.2, which shows an EIS that is interfaced to a pool of customers and a pool of resources. Both interfaces are dynamic since customers can start or end the business relationship with an enterprise, existing resources can become unavailable, and new resources can be added into the system. This EIS includes modules for scheduling and resource provisioning, data-driven learning, exception handling, customer social network, resource inventory and real-time decision making. An example scenario of dynamic adaptation is highlighted with the dotted line. An order from customer 2 is accepted by the enterprise. The order is then scheduled and allocated

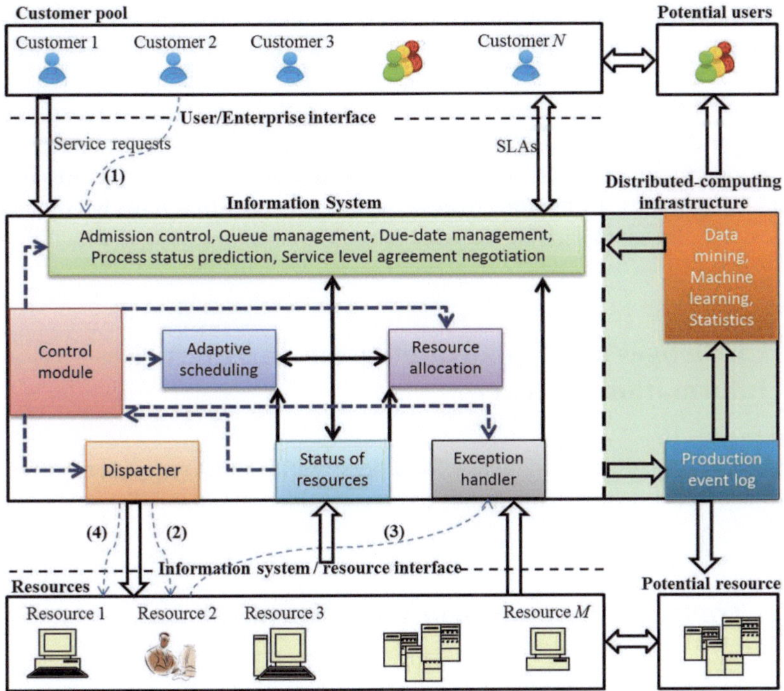

Fig. 1.2 The overall vision of this research. Research components and key technologies are elaborated

(Step 1). Certain processes of the order are allocated to Resource 2 (Step 2). However, Resource 2 becomes temporarily unavailable (e.g., due to machine breakdown, Step 3) and the order must be re-routed to Resource 1 (Step 4). All these real-time decisions taken by the EIS are hidden from the customer.

Commercial digital-print service is an example of a modern-day manufacturing-service enterprise that is primed for innovation. In the United States, the commercial print industry includes 30,000 companies and provides more than 500,000 jobs [7, 8]. Commercial print service traditionally produced homogeneous print products (e.g., thousands of copies of the same magazines) but today it is facing a paradigm shift towards mass customization and personalization owing to the digital content explosion. This paradigm shift translates into many small-sized diverse orders that demand almost instant fulfillment. This book considers digital print as a sample enterprise for the evaluation of optimization methods.

The data-driven algorithms and models proposed for an EIS in this book will enable a transformation for enterprises by supporting dynamic integration and deployment of resources, graceful recovery from failures, scalability, and dynamic adaptation without any disruption in operations. These techniques were first developed using knowledge of the system's mathematical model, and were subsequently and iteratively improved upon using reinforcement learning and optimization algorithms. The ability to carry out data-driven, online, and real-time optimization will have a profound impact on the quality of service provided by an enterprise.

The rest of this chapter is organized as follows. Section 1.1 presents an overview of the major challenges and opportunities for an EIS. Section 1.2 introduces digital-print service as an illustrative enterprise and discusses state-of-the-art solutions. Section 1.3 presents a literature review of the problems that have been tackled in the book. Finally, an outline of the book is presented in Sect. 1.4.

1.1 Challenges and Opportunities for an Enterprise Information System

In this section, we discuss challenges and opportunities that require and drive high-performance data-driven applications to be embedded into the EIS.

1.1.1 Transient, Heterogeneous and Stochastic Nature

The nature of an enterprise can be described as transient, heterogeneous, and stochastic due to the properties and the architecture of system components [9]. Heterogeneous components involve IT control and management components, error-prone processes, repetitive manual work, diverse software and hardware resources, and so on [10]. Stochasticity is another significant feature. Stochastic trends of demand and supply, and the heterogeneous nature of customer requests all drive an EIS to implement the most adaptive data-driven strategies.

Let us take the print industry as an example. Print production activities exhibit a considerable amount of stochasticity [8]. In general, continuous order acceptance and job release activities cause the workflow in the factory to be highly variable. Products that fail during manufacturing must be rescheduled for reprocessing. If resource utilization is not well-balanced by making the workload balanced among the same type of resources, then resource overloading or idling becomes a serious problem, which can be hard to track and predict.

Therefore, an EIS must be able to handle optimization problems such as task scheduling and resource allocation, and these problems should be considered in a dynamic, transient and stochastic environment. Such challenge always requires optimization solutions to be robust enough to tackle variations rather than be effective only for the steady state. In addition to being driven by up-to-date situations, data-driven strategies must be guided by historical performance and even assisted by near-future predictions. It is essential for adaptive scheduling and resource allocation to be responsive to highly varying workloads from a large amount of customers.

1.1.2   Real-Time Decision Making

The rise of the internet helps to drive various enterprise businesses to become global business. To survive, an EIS needs to be quick and flexible, which makes response time the decisive and distinguishing factor for competitiveness. Market trends and enterprise service-level performance need to be identified and even predicted in real-time under the condition that the status of resources can be determined without any delay. Real-time decision making is needed in many production management aspects such as modules for scheduling and resource allocation, data-driven learning from multi-dimensional correlated data, order admission and exception handling. In summary, processes that influence other processes should be optimized in real-time [11, 12].

For instance, in the print industry, print customers always expect delivery as soon as possible. Print service providers (PSPs) also compete for the ability to fulfill orders in the least amount of time. Facing such demand from customers, PSPs have been put in a tight squeeze in trying to meet the increased demand for variety with a shorter time-span to develop, produce, and deliver print products. The EIS of a PSP always needs to make real-time scheduling decisions such that acquired orders can be completed within the shortest possible time. Time-consuming scheduling methods might delay the manufacturing process of orders and also cause extra storage expense. Once an order contract has been established, the PSP should be able to achieve an optimized real-time schedule for this order. PSPs must not hesitate in responding to proposed contracts from customers or dealing with exceptional events that are affecting current manufacturing activities.

In summary, an EIS must be able to perform real-time decision-making in order to enable its business to be successful in the marketplace over a longer term [13].

1.1.3   Diverse and Multi-dimensional Big Data

Comprehensive data warehouses, which integrate operational data from the customer, supplier, and market data, have resulted in an explosion of information for an EIS [14]. Nowadays, enterprises have valuable information stored in a number of systems and formats such as multiple relational databases, files, web pages, packaged applications, and so on [15]. In addition to large amounts of data, big data also means interactions between data from multiple sources, and even temporal correlations.

In the print industry, massive datasets are continuously being generated on the factory floor. The EIS of a PSP has its own data warehouse in order to store all kinds of information about the PSP and its customers. For example, a large quantity of order information, such as customer information, order attributes, and order completion status, are recorded. Similarly, the enterprise event log records all major activities that take place in an enterprise, such as transactional activities, transfers of products through different departments, and exceptions. With the availability of such valuable data, the EIS is now looking for data-driven decision-making techniques that can derive insights from the available data. Data-driven approaches provide the advantage of not needing extensive expertise on production process modeling [16].

Knowledge discovery is expected to provide a high degree of integration while retaining flexibility [3]. Although data is being generated at tremendous speed, enterprises see new opportunities brought by recent advances in technologies that can help them in developing advanced solutions. Rapid improvements in computing power make it possible for an EIS to solve hard computational problems, which until recently used to be infeasible in a real-time environment. For example, the emergence of cloud computing offers enterprises creative ways to utilize IT resources [17, 18]. It allows optimization engineers to focus more on optimization objectives rather than be concerned about storage and memory limitations and insufficient computing power. Distributed computing also provides efficient solutions for an EIS to tackle hard computational problems [19, 20].

1.2 Introduction to Digital Print Production

Digital transformation of commercial print production brings new opportunities, including exploitation of embedded sensing and computing power, that enable real-time communication during various phases of the production chain and dynamic reconfiguration of the production flow [21]. Commercial print is a highly competitive service and quality-oriented business with annual retail sales of over US$700B [7, 8]. Establishing a superior service engagement and fulfillment between clients and PSPs is key to creating value and ensuring profit for a commercial print business [22]. Figure 1.3 shows the end-to-end fulfillment process of digital print services. The starting point is the engagement with clients to obtain print services [23]. The clients, who are also the content suppliers, request print services and supply content for print. Clients range from traditional publishing agencies for newspapers, books and periodicals to web-based digital media companies, e.g., *www.snapfish.com*. The PSP offers its manufacturing capability to the content suppliers as a form of utility service in exchange for payment. Today, on-demand digital print service is leading the digitization of commercial print and commanding an impressive 8 % annual growth rate [7]. Compared with traditional print services, which produce homogeneous print products (thousands of copies of the same content, also known as "long run-length"), on-demand digital print deals with high-frequency and high-volatility service demands [24].

In the commercial digital print industry, customers can request print products using multiple methods. They may physically visit the PSP or electronically submit print content to the PSP through the internet. An order is a contract for print service assigned to a PSP. It includes an order identifier, due date, number of copies, price and payment information, customer information, inks and presses, finishing and shipping methods. Several printing processes such as Raster Imaging Process

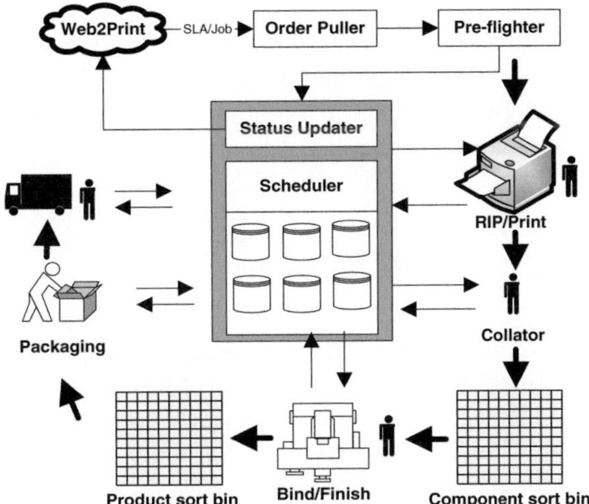

Fig. 1.3 End-to-end fulfillment process of on-demand digital print (Figure courtesy Jun Zeng [23])

(RIP), cover printing, page printing, folding, collating, binding, packaging and other related processing stages must be scheduled and finished before the completed order can be shipped to customers [25]. Besides offering high quality and good price, the PSP must deliver an order by its deadline since late delivery causes extra shipping cost and reduced customer satisfaction and/or loyalty.

In this section, we introduce the key features of digital printing production. Although illustrated through PSPs, these features are general and they exist in many enterprises.

1.2.1   Manual and Automated Rule-Based Scheduling and Resource Allocation

An enterprise that provides mass-customization services has some unique characteristics [26]. From the product's perspective, it simultaneously demands both product diversity and mass production; from an investment perspective, it is both capital-intensive and labor-intensive [8]. For instance, production scheduling on the basis of human intelligence is still the dominant practice in today's commercial print industry [23]. Every morning, the production manager examines all the print orders that need to be fulfilled that day. Based on order attributes and corresponding production plans, the production manager then sketches the product priority and the assignment of the products to each department or resource. The production departments then operate according to such an arrangement. If an exception event occurs, the production departments report to the production manager, and then the production manager updates the production schedule on the fly. Rule-based scheduling such as first-come first-served (FCFS) and resource allocation strategy such as shortest queue are still then most-popular techniques [27].

These inefficient production solutions cannot deal with high volume and a diverse production environment. Such artisanal (or craftsman-based) and static decision-making practice have been identified as one of the key reasons that the productivity growth of the print industry as a whole is falling far behind that of other manufacturing industries [28].

1.2.2 Off-line Solutions

Scheduling problems applicable to digital print production have been tackled in operations research (OR) under the framework of job-shop scheduling. For example, mixed integer linear programming (MILP) has been applied to model such problems [29]. This approach provides optimized results; however, it takes a long time to obtain optimal solutions. Therefore, these techniques cannot be used in real-time for a high-volume production environment [30]. Similarly, process-simulation tools developed to simulate the print factory system also suffer from high computational complexity, which limit their use for real-time scenarios [7, 23]. Such tools are mainly used for off-line testing of proposed solutions before they are adopted on the factory floor.

On-line scheduling solutions in operations research are based on repeatedly applying classical optimization algorithms (for instance, the aforementioned MILP) that are concerned with static input, static constraints, and static solutions. The optimization routines have an unrestricted start time when the scheduler is called. In an on-line application, this approach limits the problem size (production throughput and complexity) because the scheduling solution must be provided within a defined time interval. In addition, exceptions in the production flow are difficult to account for in such a static optimization approach [31].

1.2.3 Manual and Template-Based Order Acquisition

Order admission is another important operation on the print factory floor. The dominant practice today is that orders are acquired by customer service representatives (CSRs) who are experienced and familiar with the history of transactions [24]. When a print buyer approaches the PSP for a potential service engagement, a CSR is assigned to this print buyer. The CSR takes the print buyer's request i.e., job intent, consults the production manager, and then presents to the print buyer a quote of pricing and lead time. This approach is time-consuming and increasingly challenged by the trend of mass customization in print production.

A few of the industry-leading PSPs today implement an automated order ingestion process based on templatized product offering and service-level agreement [25]. For example, Reischling Press, Inc (RPI) spends several months to work with a strategic print buyer to develop a set of print products and rules to calculate life

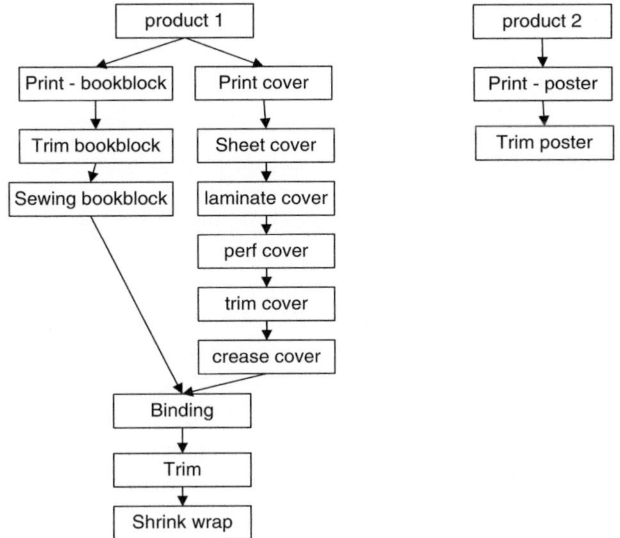

Fig. 1.4 Examples of production templates

cycles of diverse products. This set of templates is developed prior to the production launch. Figure 1.4 illustrates the production templates for two different types of products. Once the production launches, the order ingestion process is automated using this set of templates. This approach provides sufficient efficiency for order ingestion but it relies on pre-determined static rules. These rules do not reflect unanticipated production events, hence they will inevitably and quickly diverge from the reality of production floor and lead to a demand-supply mismatch. Once the mismatch grows, such methods become extremely error-prone.

1.2.4   Lack of Service-Level Forecasting and Capacity Planning

The lack of service-level forecasting and capacity planning tools can greatly influence the profitability of an enterprise [13]. Service-level forecasting and capacity planning are quite challenging in commercial and industrial print service due to the lack of useful and useable predictive tools. Consequently PSPs postpone large capital equipment purchases (e.g., million-dollar printing presses) until they observe that production at the factory floor cannot sustain the agreed service-level quality. This has a severe negative business impact on customers as PSPs either have to turn orders away or be penalized by substantial late-service charges and the cost of print buyers' goodwill [32].

1.3 Review of State-of-the-Art

1.3.1 Simulation

Enterprise simulation-based modeling can help to exploit optimization opportunities at both strategic and operational levels for an EIS [33]. It has been widely used for applications to design runtime policies. Simulation also provides performance evaluation of a system design specified by a given set of design parameters [34]. It has also been integrated with reinforcement learning to explore the significance of strategic decision making for enterprises. Simulation is very effective in helping to distinguish between different design alternatives [35].

A survey of the application of discrete-event simulation modeling to health-care clinics and systems of clinics is presented in [36]. Research on semiconductor enterprises were also studied through discrete-event simulation [37]. Applications of enterprise simulation in transportation, urban operations, supply chain management, and entertainment are described in [38].

1.3.2 Operation Optimization

Operation optimization is the process of selecting an optimal production plan and assigning limited resources for specific time periods to the set of both parallel and sequential processes for certain objectives under some practical constraints. The well-known static job-shop scheduling problems are typically NP-hard. For example, minimum makespan problems are impossible to find an optimal solution without the use of an enumerative algorithm [39].

Classical optimization techniques provide elegant static solutions to scheduling and resource management problems in an EIS [1, 40, 41]. However, these techniques are limited in their ability to handle high variability of incoming requests, dynamics of the demand and supply sides, especially when user requests are heterogeneous in nature, and when the availability of services from a large number of resources is difficult to predict [40]. For example, enterprise workflow optimization has been studied in [42] and many approaches are suggested, but they are not adaptive to the increasing trend of data-driven decision-making patterns. Researchers proposed algorithms in [43] to address scheduling and resource allocation for heterogeneous sets of machines in an enterprise; however, the problem of complex process dependencies were not considered. Point solutions to specific scheduling and allocation problems exist in the literature, e.g., adaptive scheduling and resource allocation techniques for wireless communication networks [44, 45], but these methods tend to be more reactive than predictive, and they are not generalizable to other domains.

Therefore, without considering uncertain real-time machine workloads (machine status information, e.g., queue length, maintenance schedule, and exceptional events) and shop floor dynamics (e.g., unforeseen demand, stochastic rework, and additional or unavailable resources), process plans may become suboptimal or even invalid at the time of execution. Therefore, dynamic real-time operation optimization has to been studied to achieve adaptive solutions based on the up-to-the-moment situations [46]. Although dynamic operation optimization is even more complex, it attempts to fill the gap between scheduling theory and scheduling practice, and generates applicable solutions in real-world environments [39].

The dynamic flow optimization problem for data-center networks has gained much attention in recent years [47]. For solving dynamic flow optimization problems, the main effort is on designing scalable network structure, wherein hierarchical vertical- and horizontal- scaling patterns are explored. The optimization problem is typically modeled in terms of network flow. Heuristics algorithms, e.g., ad hoc scheduling, network flow, and meta-heuristics, e.g., simulated annealing have been used to solve this problem [48–52].

The operation-optimization problem in an EIS is more complex than in a data-center network mainly due to the following reasons. First, a data-center network typically contains thousands of homogeneous clusters and a task can be assigned to any leaf cluster. In contrast, resources and tasks in an EIS are heterogeneous, introducing more complexity in task scheduling and resource allocation. Second, a data-center network has a tree topology with two or three levels of switches or routers. However, an enterprise production workflow usually involve multiple series or parallel processes. Third, a failed task in a data-center network can be re-assigned to any available cluster for rework, whereas an ad hoc rework policy has to be designed and executed depending on the specific production workflow.

As deterministic algorithms are not available for solving dynamic operation optimization problems, techniques based on heuristics (problem-specific scheduling methods), meta-heuristics (Tabu search, simulated annealing, and genetic algorithms), artificial intelligence (knowledge-based systems, fuzzy logic, Petri nets, and neural networks), and multi-agent systems (centralized and hierarchical models) have been studied [39]. Results indicate that a multi-agent approach that uses cooperative intelligent agents to develop global optimal solutions is more advantageous over traditional methods. Rather than exploring a standard architecture, multi-agent approaches consider specific challenges in a production environment and design of ad hoc strategies to achieve the best performance [46]. In addition, combining different techniques such as operational research and artificial intelligence in a multi-agent approach achieves real-time flexibility and robustness in a dynamic scheduling systems [39].

In summary, the major challenge for any EIS is to satisfy service-level agreements (SLAs), provide quality of service guarantees, and ensure profitability for providers. A transformative optimization approach is needed to analyze, predict, and decide in real-time such that the EIS provides utility to all parties.

1.3.3 Knowledge Discovery

Enterprise knowledge management/discovery is gaining popularity due to the increases in data quantity and dimensionality, as well as the rapid development in computing and storage techniques [53]. For instance, powerful and online analytical-processing tools in the EIS rely on data warehousing and multidimensional knowledge discovery. Capturing knowledge from different perspectives, and adaptively correlating and learning from them are the main directions that enterprises have been exploring in order to be more competitive [54].

The general goal in knowledge discovery for an EIS is to combine databases and data warehousing with algorithms from machine learning and methods from statistics to gain knowledge from data [3]. A review of applications of data mining in manufacturing enterprises, in particular production processes, operations, fault detection, maintenance, decision support, and product quality improvement is presented in [55]. Although significant achievements have been made in this area, there is still need for breakthroughs. For example, in the EIS decision-support system, research has shown the importance of using corporate data to derive and create higher-level information and knowledge, and the potential of combining different types of data [56]. Another motivation for further research is that machine-learning-based scheduling techniques in the EIS [57] are too simplistic to be generalizable to a distributed enterprise, especially since they tend to be offline solutions and they do not consider the larger problem of adaptive scheduling and policy management.

1.4 Outline of Book

This book research addresses a number of relevant operation-optimization and knowledge discovery problems with emphasis on data-driven, adaptive, and real-time functionality. The research topics cover different enterprise management issues, e.g., system simulation, job scheduling and resource allocation, process-execution time prediction, order-fulfillment status prediction, order-admission policy optimization, and service-level performance forecasting. Designed for different optimization purposes and implemented by different techniques, these optimization components are combined to improve the efficiency of the enterprise for changing workload and resource-availability situations. Our algorithmic innovations have been tested, validated, and demonstrated using RPI's real-life data. Developed through generalized models, these applications can be transferred to similar enterprise-software systems that need to address similar issues. The unified operation optimization and knowledge discovery framework is illustrated in Fig. 1.5.

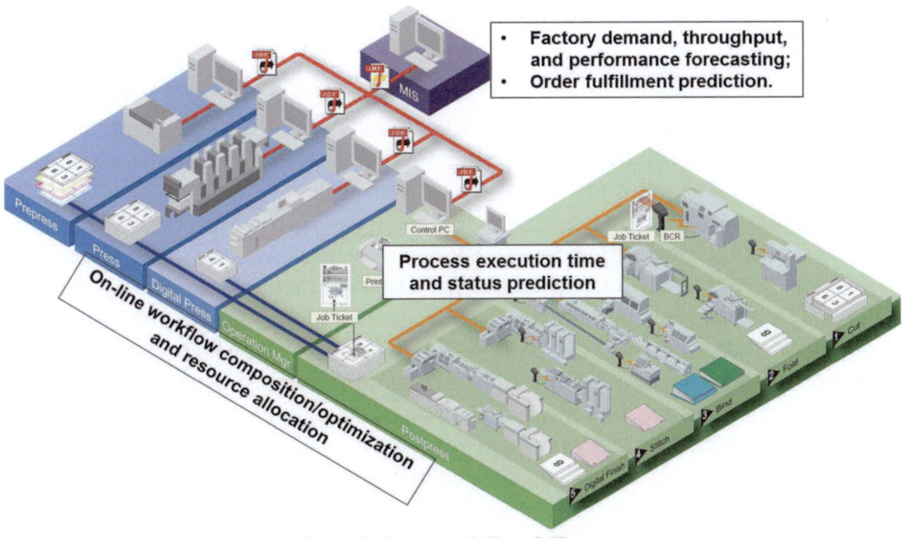

Fig. 1.5 An overview of book research

1.4.1 Simulation

Considering the heterogeneous, concurrent, and integrated characteristics of an enterprise, the first step is to adopt a holistic simulation-based approach that analyzes the design and management of an enterprise as an integrated system. Chapter 2 presents our production simulation platform built through a discrete-event simulation tool. Section 2.2 introduces stochastic discrete-event simulation method and the tool we adopted. Section 2.3 presents the virtual factory simulation platform, which mimics enterprise production activities as shown in Fig. 1.5.

1.4.2 Operation Optimization

An incremental evolutionary-algorithm based production scheduler which is used for workflow optimization is presented in Chap. 3. The background and motivation for workflow optimization is discussed in Sect. 3.1. Section 3.2 provides a formal problem description. Sections 3.3–3.7 describe the structure of the production scheduler and the underlying algorithms. Section 3.8 presents and analyzes experimental results. Section 3.9 concludes the chapter.

1.4.3 Knowledge Discovery

The remaining chapters present our online learning algorithms for hierarchical components in an enterprise. Using statistical analysis, data mining, and machine-learning techniques, we have designed and integrated new algorithms to improve the performance compared to state-of-the-art solutions.

In Chap. 4, we present a solution for enterprise process-execution time and process-execution status predictions. Section 4.1 discussed the background and motivation. Section 4.2 introduces the proposed prediction framework, problem statement, and the real-life data source used in the chapter. Section 4.3 describes in detail the state-of-the-art, the proposed time-prediction method, and the performance of method. Section 4.4 describes in detail the state-of-the-art, the proposed status-prediction method, and the performance of method. Finally, Sect. 4.5 presents conclusions and outlines ideas for further work.

Chapter 5 addresses order-admission policy optimization for enterprises. Section 5.1 discusses prior art and motivation. An order-fulfillment prediction model which includes multiple classification models has been built to estimate order life cycle. The main components and the structure of the model are discussed in Sect. 5.2. Section 5.3 describes decision-integration strategies for increasing the prediction accuracy of the model. In Sect. 5.4, we present experimental results and discuss the performance of the model. Section 5.5 concludes the chapter.

In Chap. 6, we address enterprise-wide service-level performance analysis and prediction. Section 6 discusses the background and motivation for enterprise service-level performance analysis and prediction. Section 6.1.4 introduces the state-of-the-art solutions and our data source. In Sect. 6.2, we present hierarchical time-series decomposition, the baseline model, and the proposed model for making mid-term time-series prediction. Following it, Sect. 6.3 presents time-series correlation analysis and the proposed multivariate short-term time-series prediction model used for short-term prediction. Section 6.4 concludes the chapter.

Finally, Chap. 7 summarizes the contributions of the book and identifies directions for future work.

References

1. L.D. Xu, Enterprise systems: state-of-the-art and future trends. IEEE Trans. Ind. Inform. **7**, 630–640 (2011)
2. J.H. Dunning, *International Production and the Multinational Enterprise (RLE International Business)*, vol. 12 (Routledge, New York, 2013)
3. J. Chattratichat, J. Darlington, Y. Guo, S. Hedvall, M. Kohler, J. Syed, An architecture for distributed enterprise data mining, in *Proceedings of the 7th International Conference on High-Performance Computing and Networking* (1999), pp. 573–582
4. J.-P.M.-Flatin, S. Znaty, J.-P. Hubaux, A survey of distributed enterprise network and systems management paradigms. J. Netw. Syst. Manage. **7**(1), 9–26 (1999)

5. C.L. Dunn, J.O. Cherrington, A.S. Hollander, E.L. Denna, *Enterprise Information Systems: A Pattern-Based Approach*, vol. 3 (McGraw-Hill/Irwin, Boston, 2005).
6. K. Beznosov, Engineering access control for distributed enterprise applications. Ph.D. dissertation, Florida International University, 2000
7. J. Zeng, S. Jackson, I. Lin, M. Gustafson, E. Gustafson, R. Mitchell, Operations simulation of on-demand digital print, in *IEEE 13th International Conference on Computer Science and Information Technology* (Springer, Berlin/Heidelberg, 2011)
8. J. Zeng, I.-J. Lin, E. Hoarau, G. Dispoto, Next-generation commercial print infrastructure: Gutenberg-Landa TCP/IP as cyber-physical system. J. Imaging Sci. Technol. **54**(1), 1–6 (2010)
9. S. Zykov, Designing patterns to support heterogeneous enterprise systems lifecycle, in *Software Engineering Conference in Russia (CEE-SECR), 2009 5th Central and Eastern European* (Microsoft, Moscow, 2009), pp. 83–88
10. K. Levi, A. Arsanjani, A goal-driven approach to enterprise component identification and specification. Commun. ACM **45**(10), 45–52 (2002)
11. A.W. Scheer, F. Abolhassan, W. Jost, *Business Process Automation: ARIS in Practice* (Springer, Berlin/Heidelberg/New York, 2004)
12. P. Ramanathan, J. Stankovic, Scheduling algorithms and operating system support for real-time systems. Proc. IEEE **81**(1), 55–67 (1994)
13. B. Azvine, Z. Cui, D. Nauck, B. Majeed, Real time business intelligence for the adaptive enterprise, in *The 8th IEEE International Conference on and Enterprise Computing, E-Commerce, and E-Services, The 3rd IEEE International Conference on E-Commerce Technology, 2006* (2006), pp. 1–29
14. C. Kleissner, Data mining for the enterprise, in *Proceedings of the Thirty-First Hawaii International Conference on System Sciences, 1998*, Hawaii vol. 7 (1998), pp. 295–304
15. J.M. Hellerstein, M. Stonebraker, R. Caccia, Independent, open enterprise data integration. IEEE Data Eng. Bull. **22**(1), 43–49 (1999)
16. J. Manyika, M. Chui, J. Bughin, B. Brown, R. Dobbs, C. Roxburgh, A.H. Byers, *Big Data: The Next Frontier for Innovation, Competition and Productivity* (McKinsey Global Institute, Washington, DC, 2001)
17. R. Buyya, J. Broberg, A. Goscinski, *Cloud Computing: Principles and Paradigms* (Wiley, New York, 2001)
18. P. Patel, A. Ranabahu, A. Sheth, Service level agreement in cloud computing, in *ACM International Conference on Object Oriented Programming Systems Languages and Applications*, Orlando (2009)
19. G.R. Andrews, *Foundations of Multithreaded, Parallel and Distributed Programming* (Addison Wesley, Reading, 2000)
20. F. Jammes, H. Smit, Service-oriented paradigms in industrial automation. IEEE Trans. Ind. Inform. **1**(1), 62–70 (2005)
21. J. Zeng, I.-J. Lin, E. Hoarau, G. Dispoto, Productivity analysis of print service providers. J. Imaging Sci. Technol. **54**(6), 1–9 (2010)
22. J. Spohrer, P.P. Maglio, J. Bailey, D. Gruhl, Steps toward a science of service systems. IEEE Comput. Soc. **40**(1), 71–77 (2007)
23. J. Zeng, I.-J. Lin, G. Dispoto, E. Hoarau, G. Beretta, On-demand digital print services: a new commercial print paradigm as an it service vertical, in *Annual SRII Global Conference* (2011), pp. 120–125
24. S. Karp, The future of print publishing and paid content, Publishing 2.0 blog. Tech. Rep. (2007), http://publishing2.com/2007/12/06/the-future-of-print-publishing-and-paid-content/.
25. H. Kipphan, *Handbook of Print Media: Technologies and Production Methods*, (Springer, New York, 2001), no. 40–422
26. G.D. Silveira, D. Borenstein, F.S. Fogliatto, Mass customization: literature review and research directions. Int. J. Prod. Econ. **72**(1), 1–13, (2001). [Online]. Available: http://www.sciencedirect.com/science/article/pii/S0925527300000797
27. R. Haupt, A survey of priority rule-based scheduling. Oper. Res. Spektr. **11**(1), 3–16 (1989)

28. S.P. Hoover, G.A. Gibson, The future of print and the digital printing revolution, in *31st International Conference on Imaging Science*, Beijing (2010)
29. C. Özgven, L. Özbakir, Y. Yavuz, Mathematical models for job-shop scheduling problems with routing and process flexibility. Appl. Math. Model. **34**, 1539–1548 (2010)
30. M. Agrawal, Q. Duan, K. Chakrabarty, J. Zeng, I.-J. Lin, G. Dispoto, Y.S. Lee, Digital print workflow optimization under due-dates, opportunity cost and resource constraints, in *IEEE International Conference on Industrial Informatics*, Caparica, Lisbon (2011)
31. B.L. Maccarthy, J. Liu, Addressing the gap in scheduling research: a review of optimization and heuristic methods in production scheduling. Int. J. Prod. Res. **31**(1), 59–79 (1993)
32. A. Tenhiälö, M. Ketokivi, Order management in the customization-responsiveness squeeze. Decis. Sci. **43**(1), 173–206 (2012)
33. J. Barton, D. Love, G. Taylor, Evaluating design implementation strategies using enterprise simulation. Int. J. Prod. Econ. **72**(3), 285–299 (2001)
34. L. Rabelo, M. Helal, A. Jones, H.-S. Min, Enterprise simulation: a hybrid system approach. Int. J. Comput. Integr. Manuf. **18**(6), 498–508 (2005)
35. C. Gopinath, J.E. Sawyer, Exploring the learning from an enterprise simulation. J. Manage. Dev. **18**(5), 477–489 (1999)
36. J.B. Jun, S.H. Jacobson, J.R. Swisher, Application of discrete-event simulation in health care clinics: a survey. J. Oper. Res. Soc. **50**(2), 109–123 (1986)
37. L. Rabelo, M. Helal, A. Jones, J. Min, Y.-J. Son, A. Deshmukh, New manufacturing modeling methodology: a hybrid approach to manufacturing enterprise simulation, in *Proceedings of the 35th Conference on Winter Simulation: Driving Innovation*, New Orleans (2003), pp. 1125–1133
38. R. Mielke, Applications for enterprise simulation. Simul. Conf. Proc. **2**(2), 1490–1495 (1999)
39. D. Ouelhadj, S. Petrovic, A survey of dynamic scheduling in manufacturing systems. J. Sched. **12**(4), 417–431 (2009)
40. A. Scheer, F. Habermann, Enterprise resource planning: making ERP a success. Commun. ACM **43**, 57–61 (2000)
41. T. Ibaraki, N. Katoh, *Resource Allocation Problems* (MIT Press, Cambridge, 1988)
42. C. Bussler, Enterprise wide workflow management. IEEE Concurr. **7**(3), 32–43 (1999)
43. J. Burge, P. Ranganathan, J. Wiener, Cost-aware scheduling for heterogeneous enterprise machines, in *2007 IEEE International Conference on Cluster Computing*, Austin (2007), pp. 481–487
44. Y.J. Zhang, K.B. Letaief, Adaptive resource allocation and scheduling in multiuser packet-based ofdm networks, in *Proceedings of the IEEE International Conference on Communications* (2004), pp. 2849–2953
45. M. Ergen, S. Coleri, P. Varaiya, Qos aware adaptive resource allocation techniques for fair scheduling in ofdma based broadband wireless access system. IEEE Trans. Broadcast. **49**, 362–370, 2003
46. W. Shen, Agent-based systems for intelligent manufacturing: a state-of-the-art survey. Knowl. Info. Syst. Int. J. **1**, 129–156 (1999)
47. M. Al-Fares, A. Loukissas, A. Vahdat, A scalable, commodity data center network architecture. SIGCOMM Comput. Commun. Rev. **38**(4) 63–74 (2008)
48. D. Kliazovich, P. Bouvry, S. Khan, Dens: data center energy-efficient network-aware scheduling, in *2010 IEEE/ACM International Conference on Green Computing and Communications (GreenCom) and International Conference on Cyber, Physical and Social Computing (CPSCom)*, Hangzhou (2010), pp. 69–75
49. M. Al-fares, S. Radhakrishnan, B. Raghavan, N. Huang, A. Vahdat, Hedera: dynamic flow scheduling for data center networks, in *Proceedings of Networked Systems Design and Implementation (NSDI) Symposium*, Boston (2010)
50. J.D. Moore, J.S. Chase, P. Ranganathan, R.K. Sharma, Making scheduling "cool": temperature-aware workload placement in data centers, in *USENIX Annual Technical Conference, General Track*, Anaheim (2005), pp. 61–75

51. Y. Song, H. Wang, Y. Li, B. Feng, Y. Sun, Multi-tiered on-demand resource scheduling for VM-based data center, in *Proceedings of the 2009 9th IEEE/ACM International Symposium on Cluster Computing and the Grid*, ser. CCGRID '09 (2009), pp. 148–155. [Online]. Available: http://dx.doi.org/10.1109/CCGRID.2009.11
52. L. Wang, G. von Laszewski, J. Dayal, X. He, A. Younge, T. Furlani, Towards thermal aware workload scheduling in a data center, in *2009 10th International Symposium on Pervasive Systems, Algorithms, and Networks (ISPAN)* (2009), pp. 116–122
53. U. Feyyad, Data mining and knowledge discovery: making sense out of data. IEEE Expert, **11**(5), 20–25 (1996)
54. D. O'Leary, Enterprise knowledge management. Computer **31**(3), 54–61 (1998)
55. J.A. Harding, A. Kusiak, M. Shahbaz, M. Srinivas, Data mining in manufacturing: a review. J. Manuf. Sci. Eng. **128**(4), 969–976 (2005)
56. N. Bolloju, M. Khalifa, E. Turban, Integrating knowledge management into enterprise environments for the next generation decision support. Decis. Support Syst. **33**(2), 163–176 (2002)
57. H. Aytug, S. Bhattacharyya, G. Koehler, J. Snowdon, A review of machine learning in scheduling. IEEE Trans. Eng. Manage. **41**(2), 165–171 (1994)

Chapter 2
Production Simulation Platform

2.1 Background and Motivation

Research on productivity improvement is gaining accelerated attention in enterprise industry in recent years. One notable example is the LDP solution by Xerox [1]. LDP is Xerox's simulation-based service solution offered through Xerox Managed Services (XMS) that helps to enhance print shop productivity. There are distinctions between our and their simulation-based approaches. In our approach, we modeled the print production as a content-driven cyber-physical system [2]. Anticipating that the pervasive sensing and computing-based knowledge discovery will enable additional design space and flexibility, we have tested our optimization and knowledge discovery components on the simulation platform.

In collaboration with several leading digital printing companies, we have developed a production simulation platform, i.e., a system simulator that enables us to study the operational management of a digital print factory in a quantitative fashion [3, 4]. We have adapted Ptolemy II from UC Berkeley [5] and applied a statistical discrete-event modeling approach.

The rest of this chapter is organized as follows. Section 2.2 introduces discrete-event simulation and the design tool we have used. Section 2.3 presents in detail the simulation platform.

2.2 Introduction to Stochastic Discrete-Event Simulation

Discrete-event simulation refers to modeling the operation of a system as a discrete sequence of events in time [6]. Each event occurs at a particular instant in time. The state of the system changes as events continuously take place [7]. This contrasts with continuous simulation in which the simulation continuously tracks

© Springer International Publishing Switzerland 2015

Q. Duan et al., *Data-Driven Optimization and Knowledge Discovery for an Enterprise Information System*, DOI 10.1007/978-3-319-18738-9_2

the system dynamics over time. Instead of simulating every time slice, discrete-event simulations can directly jump in time from one event to the next, therefore, they can typically run much faster than the corresponding continuous simulation. In stochastic simulation, the occurrence of an event can be stochastic and approximated by its underlying probability distribution [8]. For instance, customer arrival activities can be approximated by a Poisson distribution.

2.2.1 Ptolemy

We chose an open-source EDA toolkit, Ptolemy [5], as our modeling framework for print enterprise. Ptolemy is a Java-based, actor-oriented modeling framework for concurrent, real-time, embedded systems. Compared to object-oriented design practice, actor-oriented design emphasizes the concurrency and communication among components. Ptolemy implements a set of well-defined models of computation (for instance, continuous time, discrete event, finite state machine) that govern the component interactions. It provides a hierarchical component assembly design environment that enables the use of heterogeneous mixtures of models of computation (e.g., hybrid and mixed-signal models). It also simulates stochastic hybrid systems by adding random behavior to continuous-time models mixed with discrete events. The print production system involves control, computation, logical and physical components, for which Ptolemy's ability of blending different computational models provides the necessary simulation infrastructure support. In addition, compared with simulation methods as discussed in [9], we are already taking advantage of Ptolemy's extensive customizability that only an open-source toolkit can offer.

2.3 Virtual Print Factory

It is very challenging to achieve high productivity in digital printing with non-homogeneous arrival of orders and varying customer requirements [1, 3]. The Ptolemy-based simulation platform was first developed in [3] for simulating commercial digital print operations. It monitors the resources (e.g., utilization, inventory, health) and traces the artifacts (e.g., time-stamping after each process, faulty rates). This information is recorded in the form of relational tables in MySQL as simulation outputs. The tables of simulation results serve as a basis for further analysis and discovery. Factory-level metrics such as throughput, order cycle time, quality of service and end-to-end cost can be obtained, and statistical properties can be derived, e.g., the likelihood of one process to be faulty, and the distribution of the interruptions of one machine; visualization tools can be applied to post-process the results tables and display the large amount of data intuitively. The simulated system accounts for the performance, efficiency, stability, and sustainability as organic system attributes. Its primary utilization was for understanding various

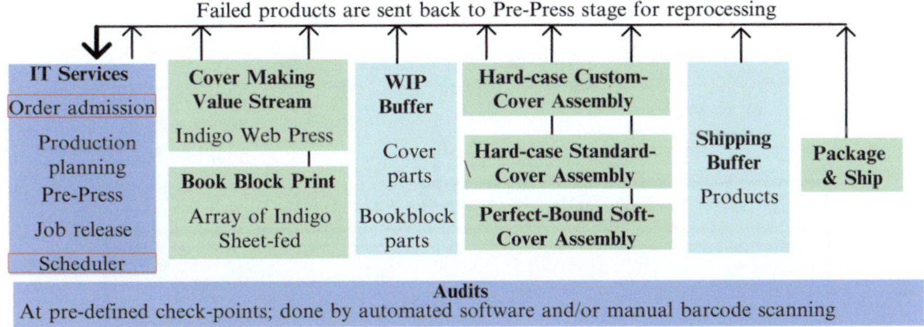

Fig. 2.1 An example of a lean manufacturing process for digital printing (Source: RPI)

lean manufacturing rules and runtime policies [10]. This platform is further used to evaluate optimization methods and decisions before their adoption on the real factory floor.

The idea of modeling document production as a manufacturing process was also described in [1]. Productivity solutions based on concepts derived from operations research were developed and deployed at several sites to accomplish substantial productivity and cost improvements for print shops [1]. The demand-driven model of "Just-In-Time" fulfillment motivates digital PSPs to emphasize innovations in lean manufacturing productivity maximization [11, 12]. RPI is an example of such a PSP whose fulfillment procedures are centered on the concept of the "value stream" [13]. Figure 2.1 shows the RPI book factory organization.

The first part is IT services shared by all value streams, including order admission, production planning, pre-press and job release. Following the pre-press, a scheduler application evaluates order priority, decomposes orders into products, and dispatches products towards different fulfillment paths. Cover making and book block printing are the following two main streamlines. After all the products of an order have arrived at the shipping buffer, this order is assembled, packaged and shipped. The end-to-end workflow is monitored at each product level by the audit function performed by either software or wireless barcode scanners via operators. Once a part of an order fails, it is sent back to the pre-press stage for reprocessing. Every processing stage may fail following a certain stochastic probability distribution that can be derived from history data.

Figure 2.2 illustrates the architecture of the operations simulation software [15] developed as a virtual factory simulation platform based on Ptolemy [5]. Figure 2.3 shows a screenshot of the virtual factory on the Ptolemy simulation platform, in which the functional blocks corresponds to the main processes shown in Fig. 2.1. The order stream is acquired from the printers' enterprise resource management (ERM) system, which encompasses both the enterprise resource planning (ERP) system and the manufacturing execution system (MES) [14]; the fulfillment paths, resources and the operating policies are programmed in the model.

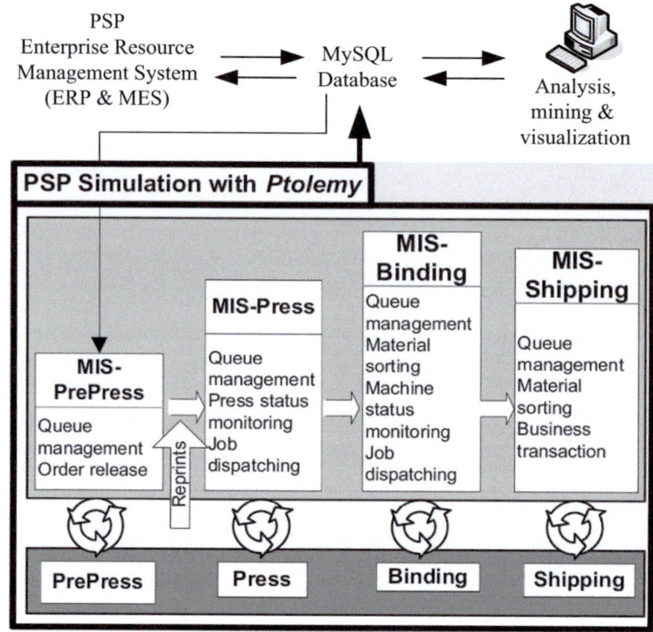

Fig. 2.2 "Virtual print factory" simulation platform architecture [14]

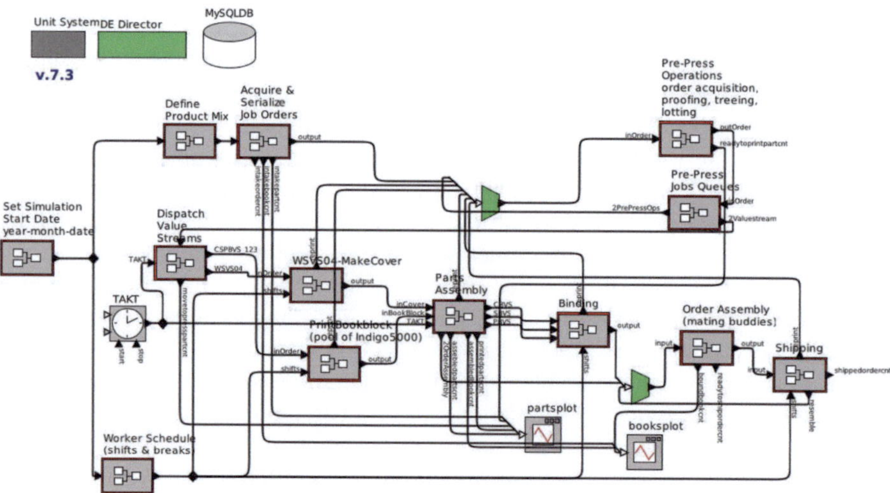

Fig. 2.3 Screenshot of the "virtual print factory" implemented on the Ptolemy simulation platform [14]

The benefit of this simulation platform is that it provides an environment that imitates the real situation on the print factory floor. Hence, we can evaluate new optimization techniques as a new scheduler application, as highlighted in Fig. 2.1 and discussed in detail in Chap. 3.

2.3.1 MySql Databases

MySQL, another open-source code, has been integrated as the database manager for all the input data and simulation results. Visual analytics tools interface with MySQL to display or extract both the factory level performance and cell-machine level dynamics. They also provide automated query and correlation discovery among different attributes.

2.3.2 Order, Product, and Part Hierarchy

An *order* is a contract established between customer and enterprise. For a PSP, the payload of an order is composed of several book titles. Each book title is referred to as a *product*. A product's information contains the digital file that will be used for preprocessing and printing, quantity (number of copies) and a fulfillment path ID. There are several types of products a factory can offer. Every type of product corresponds to its own *fulfillment path*, which defines a sequence of tasks or processing steps that completes the creation of a product of this type. A product may be further composed of more than one *part*. In case of multi-part products, the corresponding fulfillment path starts with parallel fabrication of the parts (e.g., printing the book cover and printing the book block), followed by part assembly into a product. After all products are fabricated, they can be merged to form a complete order. Based on these concepts, a processing stage can be categorized into an order-level, a product-level or a part-level stage [16].

Figure 2.4 shows an example of two simplified fulfillment paths. In the figure, we look at two different products within one order and they have fulfillment path A and fulfillment path B, respectively. Fulfillment path A has two parts, one part is used for printing cover and the second part for printing a book block. These two parts join together at the *Binding* stage to indicate the end of part processing. These two parts can be processed in parallel and are independent of each other, while *Binding* stage requires both parts to be completed before the *Binding* task of a product begins. Binding is therefore a product-level processing stage. The fulfillment path B processes a single-part product. The order-level stages *Serializing* and *Packaging* are common to all products.

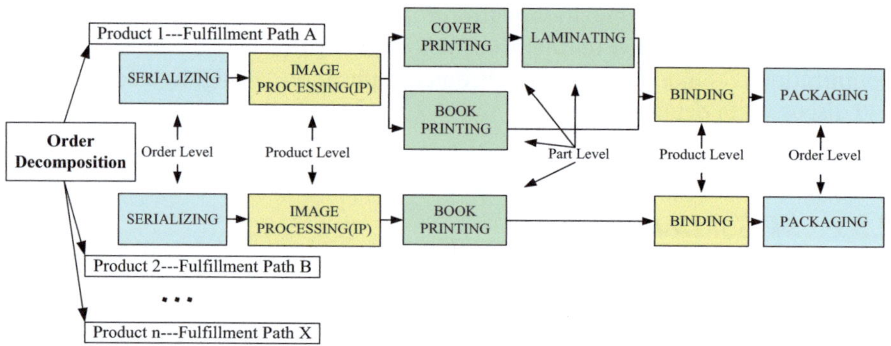

Fig. 2.4 An illustration of fulfillment paths

Table 2.1 An example of a resource library

Resource set	Quantity	Speed (mean)
Serializers	3	35
Image processors	2	20
Book printers	1	1 s/sheet
Cover printers	1	5
Laminators	1	50
Cutters	1	3
Creasers	1	3
Binders	1	42
Packagers	3	23

2.3.3 Resource Set and Task Set

The fulfillment of an order involves the processing of a sequence of steps that are serviced by different types of resources. Each resource is characterized by its functionality and speed. For example, a *color printer*'s printing rate is 1 *second per page* or a *binder*'s average service time is 42 s to bind two parts.

Based on the capabilities of each resource, the entire resource pool is partitioned into different resource sets. Resource sets for the fulfillment path example in Fig. 2.4 are shown in Table 2.1. Resources within the same resource set can process the same type of tasks; these tasks compete with each other for resource allocation. We define a task set as a set consisting of similar tasks competing with one another for the resources belonging to the same resource set. All tasks in the system can be partitioned into multiple task sets. As shown in Fig. 2.5, the tasks in the same task set are interconnected by resource contention edges. There is a one-to-one correspondence between task sets and resource sets.

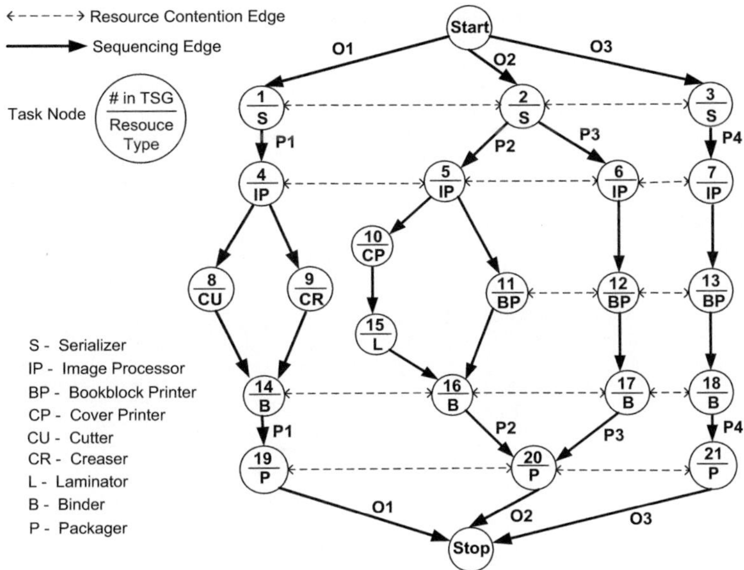

Fig. 2.5 An example of a task sequencing graph

2.3.4   Successive Order Acceptance

In a real factory, orders are accepted in a dynamic and heterogeneous manner. The VPF uses the real-life order arrival data. The state of the VPF under a normal incoming order frequency and quantity is illustrated in Fig. 2.6. The horizontal axis shows the factory time, and the vertical axis shows the number of orders accepted at 15-min intervals.

2.3.5   Stochastic Product Reprocessing

All resources are subject to stochastic malfunctions caused by various failure modes on the factory floor. It results in reworking of some products. Due to the unrecoverable nature of physical print artifacts, no matter at which processing stage a product fails, this product is sent back to the pre-press stage in order to be reprocessed from the beginning as shown in Fig. 2.1. These activities are implemented in the virtual factory as well, so that, once a product fails at any processing stage, its current manufacturing process is aborted and it is sent back to the pre-press stage immediately. If one or more products in a multi-product order require reworking, all other good products have to be stored and wait until all the reprocessed products are ready to be assembled into one order.

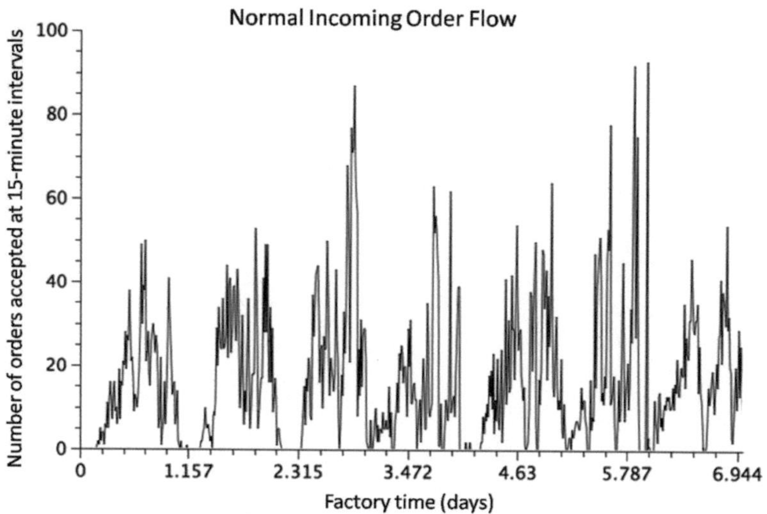

Fig. 2.6 An example of normal incoming order flow (Source: RPI real-life data in year 2009)

2.3.6 Simulation Validation

The accuracy of the simulation relies on the precision of component models included in the simulation. The fidelity of our VPF is measured by how closely it resembles reality, i.e., agreement with experimental measurements from real-life system provides highest confidence in the simulations.

The end-to-end fulfillment processes implemented and order data utilized in the VPF were directly pulled from RPI's enterprise resource planning database, and integrated into the virtual factory. The results generated by the simulation platform with RPI's internal factory audit data were validated and found to have good correlation with the situation on the real factory floor [14].

References

1. S. Rai, C.B. Duke, V. Lowe, C. Quan-Trotter, T. Scheermesser, LDP lean document production – or enhanced productivity improvements for the print industry. Interfaces **39**(1), 69–90 (2009)
2. L. Edward, Cyber physical systems: design challenges, University of California, Berkeley Technical Report No. UCB/EECS-2008-8, Tech. Rep., 2008
3. J. Zeng, S. Jackson, I. Lin, M. Gustafson, E. Gustafson, R. Mitchell, Operations simulation of on-demand digital print, in *IEEE 13th International Conference on Computer Science and Information Technology* (2011)
4. J. Zeng, I.-J. Lin, E. Hoarau, G. Dispoto, Next-generation commercial print infrastructure: Gutenberg-Landa TCP/IP as cyber-physical system. J. Imaging Sci. Technol. **54**(1), 1–6 (2010)

5. J. Davis, C. Hylands, B. Kienhuis, E.A. Lee, J. Liu, X. Liu, L. Muliadi, S. Neuendorffer, J. Tsay, B. Vogel, Y. Xiong, *Heterogeneous Concurrent Modeling and Design in Java* (EECS, University of California, Berkeley, 2001). http://ptolemy.eecs.berkeley.edu/publications/papers/01/HMAD/
6. A.I. Wasserman, Distributed discrete-event simulation. ACM Comput. Surv. **18**(1), 39–65 (1986)
7. S. Robinson, *Simulation: The practice of model development and use*, 2nd edn. (Wiley, Hoboken/Chichester, 2004)
8. B. Ripley, *Stochastic Simulation* (Wiley, New York, 1987)
9. F. Kubota, S. Sato, M. Nakano, Enterprise modeling and simulation platform integrating manufacturing system design and supply chain, in *1999 IEEE International Conference on Systems, Man, and Cybernetics, 1999. IEEE SMC '99 Conference Proceedings*, Toyko, vol. 4, no. 4 (1999), pp. 511–515
10. W. Hopp, M. Spearman, *Factory Physics*, 3rd edn. (McGraw-Hill/Irwin, New York, 2007)
11. M. Holweg, The genealogy of lean production. J. Oper. Manage. **25**(2), 420–437 (2007)
12. R. Anupindi, S. Chopra, S. Deshmukh, J.V. Mieghem, E. Zemel, *Managing Business Process Flows: Principles of Operations Management*, 2nd edn. (Prentice Hall, Upper Saddle River, 2005)
13. M. Nash, S.R. Poling, *Mapping the Total Value Stream: A Comprehensive Guide for Production and Transactional Processes*, 2nd edn. (Productivity Press, New York, 2008)
14. J. Zeng, I.-J. Lin, G. Dispoto, E. Hoarau, G. Beretta, On-demand digital print services: a new commercial print paradigm as an it service vertical, in *Annual SRII Global Conference*, Silicon Valley (2011), pp. 120–125
15. J. Zeng, I.-J. Lin, E. Hoarau, G. Dispoto, Productivity analysis of print service providers. J. Imaging Sci. Technol. **54**(6), 1–9 (2010)
16. M. Agrawal, K. Chakrabarty, J. Zeng, I.-J. Lin, G. Dispoto, Simultaneous task scheduling and resource binding for digital print automation, in *Institute of Industrial Engineers (IIE) Industrial Engineering Research Conference*, Reno (2011)

Chapter 3
Production Workflow Optimization

Operation optimization is a major function of an EIS. We present a high-performance and real-time production scheduler for an enterprise based on a dynamic incremental evolutionary algorithm. The optimization objective is to prioritize the dispatching sequence of orders and balance resource utilization. The scheduler is scalable for realistic problem instances and it provides solutions quickly for diverse products that require complex fulfillment procedures. Furthermore, it dynamically ingests the transient state of the enterprise, such as process information and resource failure probability in the production; therefore, it minimizes the management-production mismatch. Discrete-event simulation results show that the production scheduler leads to a higher and more stable order on-time delivery ratio compared to a rule-based heuristic. Its beneficial attributes collectively contribute to the reduction or elimination of the shortcomings that are inherent in today's enterprise production environment and help to enhance an enterprise's productivity and profitability.

This chapter addresses operation optimization in an EIS using digital-print industry as an example. The rest of this chapter is organized as follows. Section 3.1 discusses the background and motivation for operation optimization. Section 3.2 provides a formal problem description. Sections 3.3–3.7 describes the structure of the production scheduler and the underlying algorithms. Section 3.8 presents and analyzes simulation results. Section 3.9 concludes this chapter.

3.1 Background and Motivation

Today's on-demand digital printing industry is characterized by the drive for mass customization and personalization from all fronts e.g., consumer demand for personalized print products and broadly adopted targeted advertisement campaign in enterprise marketing. These trends are leading to a dramatic increase in the

© Springer International Publishing Switzerland 2015

Q. Duan et al., *Data-Driven Optimization and Knowledge Discovery*
for an Enterprise Information System, DOI 10.1007/978-3-319-18738-9_3

number of orders, yet the average size of an order is shrinking significantly [1]. In addition, print buyers are no longer willing to wait long for delivery. This scenario puts print service providers (PSPs) in a tight squeeze in trying to meet the increased demand for variety with a shorter time span to develop, produce, and deliver print products. Furthermore, print production is a "noisy" process in that there are high occurrences of exception events that are outside of the normal production planning process. Examples include rework resulting from the high failure probability of the production process and repair of the digital content due to file inconsistency or inadequate compliance [2]. In addition, precedence constraints between process steps within each order must be considered while scheduling. Precedence constraints increase the computational complexity of a scheduling problem [3].

Delivering orders on time is the primary goal of a PSP. If the fulfillment of an order takes longer than planned, the PSPs may incur several penalties. Upgrading shipping methods is one of the most common remedies for PSPs when an order is completed later than planned. This shipping cost upgrade can be significant; it can reach US$0.5M through a busy season [2]. The impact on goodwill is another type of penalty [4]. Printing is often a one step of a larger enterprise business process (e.g., a marketing campaign). If the printing step incurs significant delay compared to the execution time promised by the PSP, it may jeopardize the enterprise's larger business objective. Finally, delays lead to the imposition of the late penalty stated in the service contract.

This customization-responsiveness squeeze, coupled with the desire of buyers for affordable prices, forces PSPs to search for software-based automation solutions for production management in order to respond to current operational situation [5–7]. The state-of-the-art in software-based automation solutions for digital-printing production management suffers from some key limitations. Algorithms based on heuristic rules, which dominate today's leading-edge practices, are fast but far from optimal. Techniques such as mixed-integer linear programming drawn from operational research are slow and cannot provide real-time solutions for production [8]. These techniques have largely been used for off-line strategic planning. In addition, exceptions in the production flow are difficult to account for in such a static optimization approach. The manual-intensive print production process is prone to faults, which are difficult to predict. Hence static optimization contributes to a mismatch between production management and the up-to-the-moment situation in a factory [9]. In print production today, such a mismatch between the production management and the factory reality results in different forms of inefficiency, for instance, larger inventories, larger work-in-progress (WIP) accumulated at the factory floor, more active ad hoc interventions of the production manager, longer execution time, and the need for shipping upgrades. Each of these inefficiencies has a significant negative impact on the factory's profitability, e.g., larger inventories and work-in-progress (WIP) tighten up capital and larger execution time affects pricing and future business [1, 5].

To overcome the limitations of conventional methods, we present an evolutionary algorithm for optimization to achieve simultaneous production prioritization, task scheduling and resource allocation. The main contributions of this chapter are listed below:

- A risk-aware order execution-time estimation model for diverse fulfillment requirements is formulated. It takes into account scheduling of orders that are yet to be dispatched, resource allocation of tasks, stochastic nature of resource malfunctions, and factory WIP workload.
- Simultaneous task scheduling and resource allocation problem is addressed in a hierarchical manner. The first (global) step is to acquire the optimal order priority and the second (local) step is to obtain detailed product priority and task resource allocation at a lower level of granularity. The computation is carried out in a distributed framework using parallel threads. Hence, it can address high volume of print orders in real-time.
- The proposed approach can handle new orders with minimal impact on the existing factory workflow. An incremental optimization strategy has been developed in the evolutionary algorithm to maximize efficiency.
- The production scheduler dynamically ingests factory process information and the resource failure probability in order to minimize the management-production mismatch.

3.2   Problem Description and Formulation

In this section, we first describe the main components of a digital print production system, their properties and inter-relationships. Following this, a risk-aware order lead-time estimation method is presented.

3.2.1   Resources, Attributes, Parameters, and Task Sequencing Graph

Each resource can be described by its processing rate and the function that the resource performs. The set $\mathcal{R}$ include all the resources in the factory. Parameter $n_{\mathcal{R}}$ denotes the number of different types of resources. Variable s denotes the resource type, $1 \leq s \leq n_{\mathcal{R}}$. The parameter R_s denotes the set of resources of type s. The jth resource in $\mathcal{R}$ is denoted as r_j.

A task is defined as any factory operation that requires a resource. It is the atomic unit of work assigned to a resource. Tasks can be operations done by software (e.g., using computers), steps for manufacturing artifacts (require resources such as printers and binders), or shipping-related operations (require packing and shipping stations).

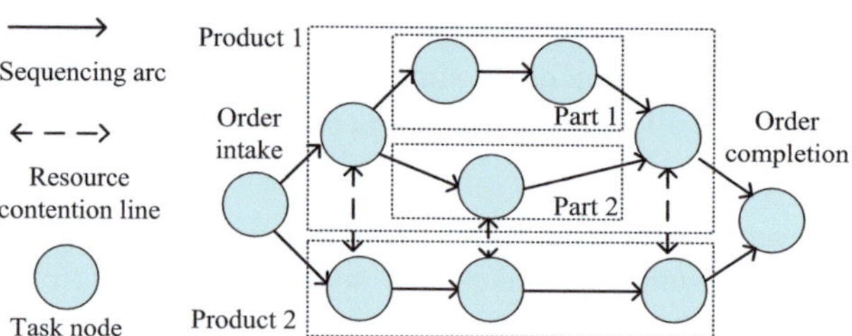

Fig. 3.1 A task sequencing graph corresponding to a single order

The fulfillment path of an order is composed of a sequence of tasks that are expanded into a series-parallel network. A task sequencing graph (TSG) is an abstract model of tasks at the system level, and it is determined by the fulfillment path and the nature of tasks. A TSG is a directed acyclic graph (DAG). Figure 3.1 shows an example of a TSG that fulfills a single order. In a TSG (V, A), each task is modeled as a vertex. The dependencies among tasks are modeled as directed arcs τ_{ij}. Arc τ_{ij} indicates that task v_i must be performed before v_j. The resource contention lines shown in Fig. 3.1 connect tasks that compete for resources in one resource set R_s.

An order is also an entity composed of several products and a product can further consist of parts. In the case of a multi-part product, parallel fabrication of parts is followed by assembly of parts into a product. In a similar manner, after the fabrication of each product, several products are merged to form a complete order [10]. The order shown in Fig. 3.1 contains two products, where the first product contains two parts.

Attribute L_i is the length of task v_i. Task length is an abstract quantity that is mapped to appropriate units based on the nature of task v_i. For example, a book printing task has the number of pages as its length unit. A binding task has the number of parts as its length unit.

Attribute M_j is the maximum length of a task that resource r_j can execute during each operation. Due to resource capacity, it cannot process a task with length greater than M_j during each operation. For example, a book printer has maximum task length of a certain amount of pages, and a binder has a maximum task length of a certain amount of parts. During the establishment of the production plan of an order, each task's length is limited by the maximum task length of the resources that can execute it.

Attribute β_j is the intrinsic failure probability of resource r_j. It depends on the resource property such life span, utilization and maintenance. It is evaluated when resource r_j executes a task with the maximum task length M_j.

Parameter α_{ij} is the failure probability of task v_i if it is executed on resource r_j. It depends on the intrinsic failure probability β_j of r_j and the task length L_i of v_i. It varies linearly with the length of v_i. It is obtained as: $\alpha_{ij} = \beta_j \times \left(L_i / M_j \right)$.

Parameter σ_{ij} is the processing rate of resource r_j when it executes task v_i. If r_j cannot execute v_i, σ_{ij} is not defined.

In our optimization framework, the production scheduler needs to estimate the total execution time (the time from entering the queue of a resource to exiting a resource after completion) of every task for each scheduling and resource allocation solution. We define functions and present steps in order to compute the total execution time of a task.

1. The first step is to assign a resource to a task. Function $\Gamma(v_i)$ returns a resource that is assigned to execute task v_i.
2. Once resource allocation is known, suppose that resource r_j is assigned to task v_i, the second step is to compute the service time. Function st_{ij} returns the service time of task v_i on resource r_j. It is obtained by dividing the length of task v_i by parameter σ_{ij}, i.e., $st_{ij} = (L_i/\sigma_{ij})$. The unit of σ_{ij} depends on the unit of L_i. If resource r_j cannot execute task v_i, then st_{ij} is not defined.
3. The third step is to compute the wait time of v_i. Since a resource executes tasks using a first-come first-served (FCFS) policy, the wait time of v_i comes from the service time of tasks in front of it. There are two types of tasks that can wait in front of v_i in the queue. One type is WIP tasks, i.e., tasks belonging to the orders that are currently being manufactured. The other type is scheduled tasks, i.e., tasks that are assigned to resource r_j by the production scheduler and belonging to the orders that have not yet entered the production process.

   In order to be executed, task v_i needs to wait until resource r_j finishes the execution of both WIP and scheduled tasks. Therefore, the wait time of v_i consists of two components. The first component qwt_j is the service time of WIP tasks. The second component swt_{ij} is the service time of scheduled tasks. The information of WIP tasks is provided by the resource management database, which records and updates resource information such as queue length. The value of qwt_j only depends on the status of resource r_j when function wt_{ij} is called. The value of swt_{ij} depends on the number and size of scheduled tasks have been mapped to resource r_j before task v_i is mapped to resource r_j by function $\Gamma(v_i)$. Therefore, different scheduling and resource allocation solutions result in different values of swt_{ij}. Function wt_{ij} returns the sum of the two components, i.e., $wt_{ij} = qwt_j + swt_{ij}$.
4. The last step is to compute the total estimated execution time, which consists of service time and wait time. Function t_{ij} returns the sum of st_{ij} and wt_{ij}, i.e., $t_{ij} = st_{ij} + wt_{ij}$.

3.2.2   Risk-Aware Execution-Time Estimation

The execution time of an order, a product or a part is defined as the time to complete the manufacturing of an order, a product or a part, respectively. The most popular execution-time computation strategies are simple static rule-based methods [1]. We describe a more rigorous method for computing an order's risk-aware execution time based on the factory up-to-the-moment status.

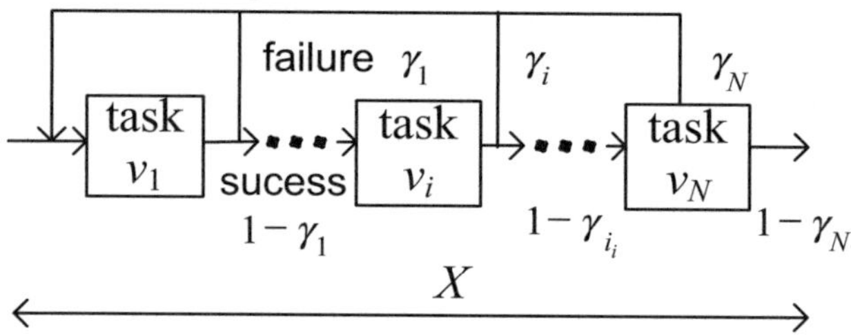

Fig. 3.2 Manufacturing process for a single-part product

In the case of a single-part product, its manufacturing process involves successful completion of a sequence of tasks, shown in Fig. 3.2. The rework policy is defined such that if any task fails, the manufacturing process restarts from the first task. Each task has a specific resource assigned to it given by function Γ. Variable δ_{ij} is a binary random variable such that $\delta_{ij} = 1$ if resource r_j is assigned to execute task v_i, $\delta_{ij} = 0$ otherwise. Parameter w_i is the time required to execute task v_i on the resource to which it is mapped. It is obtained using Eq. (3.1).

$$w_i = \sum_{j=1}^{|\mathcal{R}|} \delta_{ij} \cdot t_{ij}, \quad \sum_{j=1}^{|\mathcal{R}|} \delta_{ij} = 1 \tag{3.1}$$

Parameter γ_i is the failure probability of task v_i on the resource that is assigned to execute it. It is obtained using Eq. (3.2).

$$\gamma_i = \sum_{j=1}^{|\mathcal{R}|} \delta_{ij} \cdot \alpha_{ij}, \quad \sum_{j=1}^{|\mathcal{R}|} \delta_{ij} = 1 \tag{3.2}$$

The constraint $\sum_{j=1}^{|\mathcal{R}|} \delta_{ij} = 1$ in Eqs. (3.1) and (3.2) ensures that each task can only have one resource assigned to it. Each task has a specific resource assigned to it given by function Γ. Parameter w_i is the time required to execute task v_i on the resource to which it is mapped. Parameter γ_i is the failure probability of task v_i on the resource that is assigned to execute it. Each task can only have one resource assigned to it. Random variable X is the execution time of a single-part product. The expectation $E[X]$ of X can be obtained by Eq. (3.3):

$$E[X] = \sum_{i=1}^{N} \frac{w_i}{\prod_{k=i}^{N} (1 - \gamma_k)}. \tag{3.3}$$

Equation (3.3) is only applicable in the case of a chain of sequential tasks such as a single-part product. Derivation and Proof of Eq. (3.3) are presented in Appendix A.

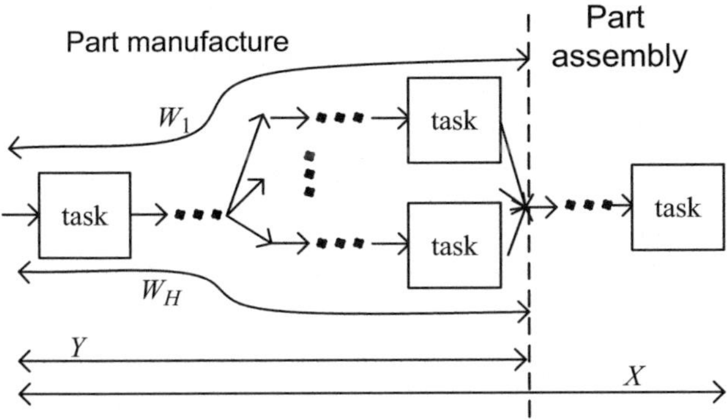

Fig. 3.3 Manufacturing process for a multi-part product

The manufacturing process for a multi-part product is shown in Fig. 3.3. Parts can be manufactured in parallel. Hence, the production process for an individual part is equivalent to a single-part product. After all the parts are manufactured, they are assembled into one product. Discrete random variable Y is the execution time for completing all the parts. Therefore, Y is defined as: $Y = \max_{j \in \{1,...,H\}}\{W_j\}$, where W_j is a discrete random variable that represents the execution time of the jth part. The expectation of random variable Y is approximated as follows to facilitate real-time computation:

$$E[Y] \approx \max_{j \in \{1,...,H\}}\{E[W_j]\} \cdot \sum_{j=1}^{H} \frac{1}{j}. \tag{3.4}$$

The expectation of Y can be obtained from its probability mass function (PMF) and cumulative density function (CDF), which requires the PMF and CDF of each W [11]. The computation time increases exponentially with the number of parts and the number of stages. Detailed derivation procedure is presented in Appendix B. From simulation results, we observed that W can be approximated as an exponential distribution with rate $1/E[W]$. Detailed approximation procedure is presented in Appendix C. The expectation of the maximum of H independent identically distributed exponential random variables with rate $1/\max\{E[W]\}$ is $\max\{E[W]\} \cdot \sum_{j=1}^{H} 1/j$ [12]. Therefore, we estimate the expectation of Y as shown in Eq. (3.4) for simplicity. The rework policy is defined such that if any part fails to merge into one product, then all the parts need reprocessing. We can view the set of tasks that manufacture all the parts of a product as a single task whose expected execution time equals $E[Y]$. After simplifying all the tasks for manufacturing all the parts as the first stage, we can compute the expectation of the execution time of a multi-part product by assuming that there are N stages in the chain. Therefore, $E[X]$ can be obtained by the following equation:

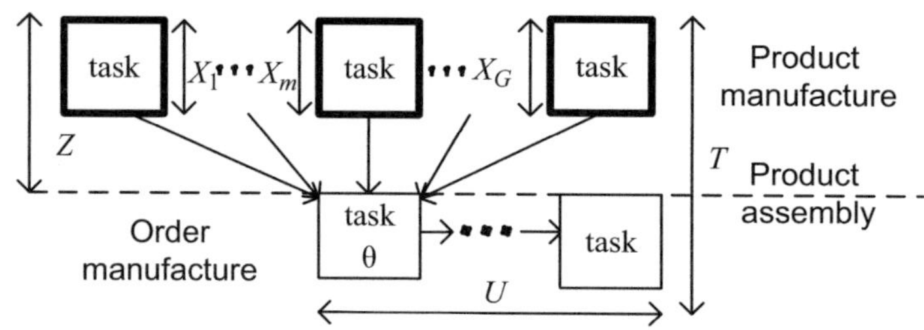

Fig. 3.4 Manufacturing process for a multi-product order

$$E[X] = \frac{E[Y]}{\prod_{k=1}^{N}(1-\gamma_k)} + \sum_{i=2}^{N} \frac{w_i}{\prod_{k=i}^{N}(1-\gamma_k)}. \tag{3.5}$$

After substituting (3.4) into (3.5), we get:

$$E[X] = \frac{\max\limits_{j\in\{1,\dots,H\}}\{E[W_j]\}\sum_{j=1}^{H}\frac{1}{j}}{\prod_{k=1}^{N}(1-\gamma_k)} + \sum_{i=2}^{N} \frac{w_i}{\prod_{k=i}^{N}(1-\gamma_k)}. \tag{3.6}$$

The manufacturing process for a multi-product order is shown in Fig. 3.4. After all the products have been manufactured, they are assembled into one product assembly. Note that, discrete random variable Z is the execution time of completing all the products of an order, i.e., the execution time of the product assembly. Therefore, Z is defined as: $Z = \max_{j\in\{1,\dots,G\}}\{X_j\}$, where X_j is the execution time of the jth product and G is the number of products in a multi-product order.

The expectation of Z is approximated as follows:

$$E[Z] \approx \max_{j\in\{1,\dots,G\}}\{E[X_j]\} \cdot \sum_{j=1}^{G}\frac{1}{j}. \tag{3.7}$$

The approximation process is similar to that of Eq. (3.4). After all the products have been manufactured, remaining tasks comprise a sequence that is applied to the product assembly. The first of the remaining tasks is referred to as task θ. It identifies defective products. The rework policy is defined such that flawed products are sent back for reprocessing from the beginning. In the meantime, qualified products wait at task θ for the reprocessed products to arrive.

In general, reprocessed products always have the highest priority in manufacturing. Print factory data shows that the failure probability of a task θ, denoted by γ_θ, is in the range of 10^{-4} to 10^{-3}, where data has been obtained from Reischling Press, Inc. (RPI) [1].

As shown in Fig. 3.4, variable U is the execution time of completing the remaining production process for an order after all the products have been manufactured.

It can be calculated using the steps described for computing the execution time of a single-part product. Variable Z can be obtained by Eq. (3.7). Variable T is the execution time of an entire order and it can be estimated using the following equation:

$$T = Z + (1 - \gamma_\theta) \cdot U + \gamma_\theta \cdot \left(\frac{\sum_{m=1}^{G} E[X_m]}{G} + U \right), \tag{3.8}$$

where $\sum_{m=1}^{G} E[X_m]/G$ is the expected time needed to reprocess the failed products. By applying the linearity property of expectation of the sum of random variables [11], we can compute the expectation of T as follows:

$$E[T] = E[Z] + (1 - \gamma_\theta) \cdot E[U] \tag{3.9}$$

$$+ \gamma_\theta \cdot \left(\frac{\sum_{m=1}^{G} E[X_m]}{G} + E[U] \right).$$

In Eq. (3.9), $E[Z]$ is computed using Eq. (3.7), $E[U]$ is computed using Eq. (3.3), and $E[X_m]$ is computed using Eqs. (3.3) or (3.6) if the mth product is a single-part product or a multi-part product, respectively.

Therefore, we can obtain the expectation of the risk-aware execution time of any single or multi-product order, with either single or multi-part products by Eq. (3.9).

3.2.3 Normalized Risk-Aware Slack

Every new order faces the risk of not being completed on time. This risk is incurred due to three factors: (i) task rework caused by stochastic machine malfunctions; (ii) influence from other new orders; (iii) manufacturing status of WIP orders. The probabilistic model that computes the expectation of the risk-aware execution time of an order comprehensively accounts for the main factors that are responsible for increasing the risk of late delivery.

The risk-aware execution time derived from Eq. (3.9) is then used to compute the slack (the temporal difference between due time and completion time). For each order, we calculate the normalized risk-aware slack as follows:

$$NS_i = \frac{d_i - t_j - E_i}{d_i - t_j} = 1 - \frac{E_i}{d_i - t_j}.$$

where E_i is the expectation of the risk-aware execution time of the ith order, d_i is the due time of the ith order. Variable t_j is the release time of this set of orders. It is the real time in the factory when factory releases this set of new orders for manufacturing at the jth _TAKTTrigger_ signal (details are explained in Sect. 3.6). Therefore, $d_i - t_j$ is the duration within which the ith order is expected to be completed.

3.3   Production Scheduler

Figure 3.5 shows the system architecture and the activity diagram of the production scheduler in a print production environment. The production scheduler consists of two main components, the *dispatcher* and the *scheduler*. They cooperatively execute a two-step production scheduling algorithm.

There are notable differences as well as similarities between the dispatcher and the scheduler. After being assigned with a production plan and a due time, an order enters the production scheduler and joins the order pool of the dispatcher, which contains orders that are yet to be dispatched for manufacturing. In the first step of creating production scheduling, the role of the dispatcher is to explore the optimal dispatching sequence for orders in the order pool. Top N orders from the optimal dispatching sequence will be sent to the scheduler. In the second step, the scheduler further determines the optimal dispatching sequence for all the products and resource allocation for every task derived from the top N orders. The value of N is dynamically determined by the changing throughput of the factory, which is further driven by the demand of customers.

This two-step incremental scheduling strategy is designed to first solve the order-level (highest granularity level) prioritization problem, then tackle the product-level (middle granularity level) prioritization and the task-level (lowest granularity level) resource allocation problem. This kind of scheduling strategy is different from traditional methods that try to achieve deterministic scheduling and/or resource

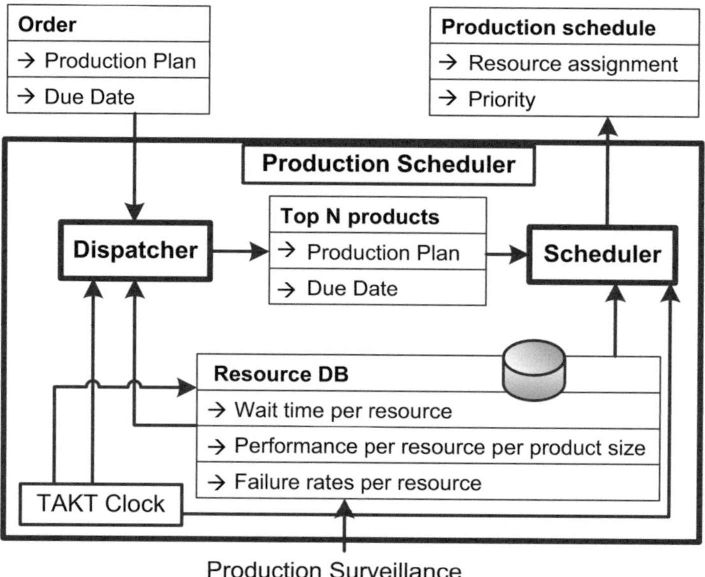

Fig. 3.5 System architecture of the production scheduler

allocation at the lowest granularity level in one step [13]. Researchers have shown that for many applications, partitioning of large optimization problems can lead to good solutions in real-time [14]. For example, Fei, Meskensa and Chu [15] addressed a planning and scheduling problem for an operating theatre in two phases and demonstrated enhanced efficiency. Note that, in a streamlined factory, it is not necessary to explore task-level prioritization. Only orders and products can be sorted and dispatched according to a priority sequence at certain stages of the manufacturing process. Tasks naturally inherit priorities from products since they are executed by resources obeying first-come first-served (FCFS) policy in a streamlined manufacturing environment. Although FCFS does not necessarily result in optimal or even superior solutions, FCFS results in fairness in terms of resources made available to incoming orders. In addition, FCFS is easy to implement. The application of FCFS can therefore be widely observed in practice [16].

3.4   Problem Complexity Analysis

The optimization problems tackled by the production scheduler are NP-hard. Heuristic algorithms are needed for practical problem instances. It has been shown in [17] that sequencing orders with random execution times to minimize the variances of completion times is an NP-complete problem. For convenience, we denote this problem by *P1*. The dispatcher attempts to minimize the sum and variance of normalized risk-aware slack times. Suppose that orders have the same due time. Our problem is denoted by *P2*. The objective function of *P1* is $obj_1 = Var(C(S))$, where S represents an order sequence, $C_i(S)$ is the completion time of order i under S. The objective function of *P2* is $obj_2 = Var(NS(S))$, where $NS_i(S)$ is the normalized slack time of order i under S. Completion time and slack time are related as follows:

$$NS(S) = \frac{DueTime - C(S)}{DueTime - ReleaseTime} = c_1 - c_2 C(S) \tag{3.10}$$

where c_1 and c_2 are positive constants. Therefore $Var(NS(S)) = Var(c_1 - c_2 C(S))$. By applying these properties of variance: $Var(X + a) = Var(X)$, and $Var(aX) = a^2 Var(X)$ [11], where a is a constant, we can get the transformation as $obj_2 = c \cdot obj_1$, where c is a positive constant, $c = 1/c_2^2$.

For any given instance of *P1*, we construct an instance of *P2* by multiplying obj_1 with the positive constant c. The resultant instance is one of *P2*. This construction takes polynomial time. A solution of the transformed problem instance minimizes obj_2, which also minimizes obj_1. Hence, this solution is also a valid solution for the initial problem instance. Conversely, a solution of the initial problem instance also minimizes obj_2, which means it is also a solution of the transformed problem instance. Therefore, we have shown that *P2* can be reduced from *P1* in polynomial time. The order-prioritizing problem that the dispatcher attempts to solve is therefore NP-hard.

The optimization problem of the scheduler is to find the optimal scheduling sequence for products as well as optimize resource allocation for tasks. The problem is subject to resource constraints that there are one-to-one mappings between resource sets and task sets. As discussed for the dispatcher, sequencing products is at least as hard as a known NP-complete problem. Several scheduling problems with resource constraints are discussed in [18, 19] and optimal resource allocation problems are discussed in [20]. They are at least as hard as known NP-complete problems. Our objective combines the two objectives. Therefore, based on the principle of restriction, it follows that the problem the production scheduler attempts to solve is computationally at least as hard as know NP-complete problems.

3.5 Incremental Genetic Algorithm

Incremental genetic algorithms (IGA) [21] are used in both the dispatcher and the scheduler as the main heuristics. The core algorithm, GA, is an evolutionary population-based meta-heuristic optimization strategy [22]. A chromosome is a sequence of numbers that encodes a proposed solution to the problem that the GA algorithm is trying to solve. Analogous to biological evolution process, reproduction, mutation, recombination and selection mechanisms are implemented in GA. The evolution process of a population is realized by repeated applications of these mechanisms so that the quality of the next population can be improved based on the previous population. A fitness function in GA evaluates a population of candidate solutions so that the best solution can be distinguished. After reaching a non-improving best solution or a computation time limit, as well as receiving a triggering signal that requires a solution, GA returns the best solution obtained so far. GA-based strategies have been recognized as efficient solutions for heuristically solving complex and intractable optimization problems across various domains, especially in scheduling problems [23–26].

Furthermore, a general characteristic has been identified in our optimization problem. Whenever an order joins or leaves the order pool of the dispatcher, impact on the optimization problem is minor, i.e., addition or removal of one order only causes usually minor changes to the optimal dispatching sequence of other orders. Considering this characteristic, an IGA is adopted in order to reduce the time cost to reoptimize the problem after each change of the order pool. IGA is similar to classical GA except that it starts with an initial population that contains chromosomes generated based on the best chromosome obtained from solving the previous problem, which corresponds to the problem prior to a change of the order pool.

In the dispatcher, the computation is carried out in a distributed framework using parallel threads of GAs. Each thread is referred to as a *dispatching GA* thread. Similarly, a GA thread in the scheduler is referred to as a *scheduling GA* thread.

3.6 Dispatcher

The dispatcher is a dynamic incremental optimization engine that is continuously running in the system. Its main function is to maintain a prioritized sequence of orders that are going to be dispatched for manufacturing. Every new order is first sent to the dispatcher.

Activities of the dispatcher are triggered by two external control signals. The first signal is a periodic signal called _TAKTTrigger_. The _TAKT_ clock in Fig. 3.5 generates a new _TAKTTrigger_ signal with a constant time interval, called _TAKT_ interval. An example value of the _TAKT_ interval is 15 min, which is the actual setting used by RPI. The second signal, called _IncomingOrder_ signal, is generated whenever a new order is sent to the dispatcher.

Whenever there is a _TAKTTrigger_ signal, the dispatcher pulls the newest resource information such as failure probability and queue length from a resource database. Data are updated in real time by the production surveillance system. Therefore, parameters that the dispatcher uses are always up-to-the-moment. Updating information is crucial for minimizing the management-production mismatch. At each _TAKTTrigger_ signal, the dispatcher also releases a prioritized list of orders to the scheduler. Simultaneously, these orders are also removed from the order pool of the dispatcher.

There can be more than one dispatching GA threads running in the dispatcher. The order pool of the dispatcher is partitioned into many smaller _buckets_ according to order due time. The partition is implemented such that all the buckets have similar quantities of orders with close due times. The _due time bound_ of each bucket is determined by the earliest and latest due times of orders in this bucket. Each bucket corresponds to a dispatching GA thread.

Whenever there is a new order, it is inserted into a bucket whose due time bound covers or is closest to its due time. Meanwhile, other buckets are not affected. When the loads of buckets become imbalanced to a certain degree, an automatic partitioning process applies to re-distribute orders into new buckets.

The dispatching GA in the host bucket of the new order inserts this order to its intermediate solutions by adding one gene to its chromosome. After modifying all chromosomes, the GA thread resumes running. In this way, the candidate solutions of the updated problem that contain the new order are built upon the intermediate solutions of the previous problem that do not contain the new order. The incremental strategy is used to accelerate the convergence to the new optimal solution. At the same time, only one dispatching GA is interrupted, whereas other GA threads are isolated from the changes resulting from the new order.

Order priority sequences obtained from buckets are then assembled into a global prioritized order list according to the due time bounds. The above methodology follows a distributed framework. The production scheduler as shown in Fig. 3.5 partitions a large computational problem over an extensive dataset into distributed sub-problems and then integrates results from sub-problems. It overcomes the difficulty that the original large problem cannot achieve a satisfactory solution in real time.

3.6.1 Scheduling Priority, Resource Allocation Policy, and Fitness Function in the Dispatching GA

In the dispatching GA, priority is assigned at order level. Therefore, products from the same order have no difference in terms of manufacturing sequence. Hence, for each resource, the execution sequence of tasks from different products of the same order can be arbitrary. However, tasks from different orders have a priority sequence decided by the priority sequence of their parent orders.

GA manipulates a set of candidate solutions in each *generation*. A solution is encoded into a *chromosome*, which is a vector of *genes*. We utilize a random key representation technique, which allows us to realize a good chromosome representation for easy and unambiguous decoding of a chromosome into a valid solution [27]. Each gene in the chromosome is a random fractional number between 0 and 1. The genes are used for encoding the priorities of orders. If the total number of orders in a dispatching GA is n, and orders are numbered from 1 to n, then each chromosome can be represented as a vector *chromosome* = (*gene*(1), ..., *gene*(m)). The priority of order i is given by *gene*(i). Another advantage of such chromosome representation is that after applying simple evolutionary techniques, such as crossover, the resulting chromosome is still a valid solution.

Crossover is a technique to exchange genes between two chromosomes. We use the parameterized uniform crossover approach in this work [28]. In mutation, the remaining part of the population is randomly created in the next generation.

In the dispatching GA, decoding each chromosome determines order priority, whereas the resource allocation information is not determined. However, in order to compute wait time and service time of each task, resource allocation information must be known. Therefore, the dispatching GA requires a good approach to estimate resource allocation such that it does not deviate too much from resource allocation results given by the scheduling GA in the following computation.

An analogy to a classical optimization problem has been identified in our work. The process of computing wait time and service time for tasks that need to be processed by resources in the same resource set R_s, $1 \leq s \leq n_{\mathcal{R}}$, is similar to the "minimum makespan" scheduling problem [29]. "Minimum makespan" scheduling problem is formulated as follows. Given a set of jobs and a set of machines, the jobs have either identical or different processing times on the given machines. The goal is to assign jobs to machines so that the maximum completion time which is referred to as makespan is minimized. The order in which the jobs are processed on a particular machine does not matter. Similarly, our problem of scheduling tasks of the same type and from the same order is formulated as follows. Given a set of tasks from the same order that use the same set of resources, a task v_i has service times t_{ij} if mapped to resource r_j, the problem is to find an assignment of tasks to resources such that the maximum completion time of tasks is minimized. Since tasks are from the same order, there are no precedence constraints between them.

One policy is as follows: (i) sequence tasks arbitrarily; (ii) schedule the next task on the resource that has been assigned the least amount of workload so far, i.e., the least accumulated service time for tasks in the resource queue. The minimum makespan scheduling theorem shows that this solution is a $(2 - 1/|R_s|)$-approximation to the optimal minimum completion time for tasks [30]. This policy is referred to as minimum makespan scheduling and resource allocation policy.

In the dispatching GA, tasks are first sequenced by order priority and then assigned resources according to the minimum makespan policy. The wait time and service time for each task is computed afterwards.

Although optimal resource allocation can only be achieved by applying task-level priority assignment, by adopting order-level priority assignment and the minimum makespan policy, the dispatching GA can still achieve a certain degree of near-optimality. Since the scheduling GA attempts to optimize the scheduling and resource allocation at the lowest granularity level, the results of the dispatching GA are at most $(2 - 1/|R_s|)$ times worse than the scheduling GA [30].

We have discussed that the dispatching GA adopts dynamic estimation of the risk-aware slack. It is neither the static estimation used by rule-based heuristics nor worst-case estimation as in our previous work [10]. It is close to the result that can be achieved by the scheduling GA. This is important since orders and products are eventually dispatched for manufacturing according to solutions given by the scheduling GA. We need to minimize the difference between the estimates of the dispatching GA and results of the scheduling GA.

The objective of the dispatching GA is to explore the optimal priority sequence of orders while honoring each order's due-date commitment, as well as accounting for stochastic factors.

A two-step fitness function evaluation framework is adopted in the dispatching GA [23]. The first fitness function is to minimize the average of the normalized risk-aware slacks $\overline{NS}$ of orders, i.e., $\overline{NS} = 1/P \sum_{i=1}^{P} NS_i$. It is formulated as follows:

$$\text{minimize} f_1 = \overline{NS} \tag{3.11}$$

where P is the number of orders in a dispatching GA. The objective of fitness function (3.11) is to meet JIT production requirement since storage of physical products on the factory floor can be expensive [31].

The second fitness function is designed to minimize the sum of the variance of normalized risk-aware slack of orders.

$$\text{minimize} f_2 = \sum \left(NS_i - \overline{NS} \right)^2 \tag{3.12}$$

The objective of fitness function (3.12) is to preserve a reasonable amount of slack for orders according to order attributes and factory status.

Each chromosome in a dispatching GA is first evaluated by fitness function f_1. If multiple chromosomes have the same fitness value according to f_1, then f_2 is further computed. Therefore, the best chromosome has the minimum values of both f_1 and f_2.

In addition, two candidate chromosomes that encode EDD (earliest due date) and min-slack (minimum slack time) scheduling solutions are included in the initial population of the dispatching GA to guarantee that the solution is at least as good as what can be derived from rule-based heuristics.

3.7 Scheduler

The scheduler contains only one scheduling GA thread. It is applied over the prioritized orders sent by the dispatcher. The activation and termination of the scheduling GA are controlled by _TAKTTrigger_ signal. Once a _TAKTTrigger_ signal arrives, current scheduling GA terminates and produces the production scheduling solution. The production scheduler dispatches these orders for manufacturing. A new scheduling GA starts running over the new set of prioritized orders sent by the dispatcher.

3.7.1 Scheduling Priority, Resource Allocation Policy, and Fitness Function in the Scheduling GA

The scheduling GA gives the final production scheduling solution, in which the priority is assigned at product level. The resource allocation for each task is also given by a scheduling solution. Based on the order priority sequence given by the dispatcher, the scheduler attempts to obtain the optimal priority sequence of products and resource allocation for each task within one *TAKT* interval.

A chromosome of the scheduling GA contains two types of information. The first part encodes the product priority, the second part encodes the resource allocation for every task. The scheduling GA has a thoroughly determined resource allocation information for every task encoded in each chromosome, therefore, the resource allocation policy followed by the scheduling GA is to directly decode a chromosome. Decoding product priority is similar to decoding order priority as discussed for the dispatching GA. Decoding resource allocation is done as follows. The resource allocation gene of task v_i is $gene(M + i)$, where M is the total number of products. If v_i is mapped to resource set R_s, then resource $\lceil gene(M + i) \times |R_s| \rceil$ in R_s is assigned to execute v_i. The length of a chromosome in the scheduling GA is equal to the sum of the number of products and the number of tasks from all orders. Therefore, chromosome length is much larger than it in the dispatching GA even if they contain the same set of orders.

The dispatcher has already optimized order-level priority, therefore, the main objective of the scheduler is to achieve an optimal resource utilization solution. Furthermore, the production scheduler is designed for high volume and tight due-date orders, when long queues at resources can easily become bottlenecks in the manufacturing system if an ineffective resource allocation policy is adopted. Resource utilization can be measured in terms of time. It is the total time that a resource is executing tasks [20].

Fair resource utilization is achieved when in each resource set R_s, the utilization of the most heavily loaded resource is close to the utilization of the least loaded resource. The fitness function of the scheduling GA is designed to minimize the sum of the normalized utilizations of the most heavily loaded resources from all the resource sets and it is formulated in Eq. (3.13).

$$\text{minimize} f_3 = \sum_{s=1}^{n_{\mathcal{R}}} \frac{\max\{u(r_j)|r_j \in R_s\}}{\sum_{r_j \in R_s} u(r_j)} \qquad (3.13)$$

After a chromosome in the scheduling GA is decoded, product priority and task resource allocation are all determined. Then risk-aware execution time is computed for each order. With all the information available, the scheduling GA computes the value of fitness function f_3. If multiple candidate chromosomes have the same minimal fitness value according to f_3, then the scheduling GA selects the best candidates according to the two-step fitness function framework of the dispatching GA, as discussed in Sect. 3.6.1.

3.8 Validation

3.8.1 Evaluation Metrics

The non-homogeneous arrival of orders is described by *order arrival rate*, which is the number of orders arrived during a certain interval. Parameter *throughput* is the number of orders delivered during a certain interval. *Instantaneous workload* is the number of WIP orders being processed at a specific timestamp. *Average workload* is the average of instantaneous workload during a certain interval. *Normalized average workload* is the average workload divided by the maximum average workload. Parameter *on-time delivery ratio* is an interval-based measure of production efficiency, which determines how efficiently a factory meets its customer's deadlines for an interval. It is defined as the ratio of the number of orders delivered on time versus the total number of orders delivered during a certain interval. Similarly, parameter *cumulative on-time delivery ratio* is the total number of orders delivered on time versus the total number of orders delivered from time zero to a given time.

3.8.2 Simulation Settings

The production scheduler has been integrated in the VPF. The rule-based heuristic used by RPI and implemented in the VPF is min-slack for product prioritization and round-robin for task resource allocation. The min-slack approach assigns priorities to product by sequencing them according to slack time estimated based on a set of fixed values. Resources in the same resource set take turns to execute tasks by the round-robin approach. All tests and simulations were run on a 32-bit Linux operating system with 2 GB of RAM.

As shown in Fig. 2.6, we considered a real-life order-arrival data from RPI which has a *TAKT* interval equals 15 min. There are 12 different product types, the number of tasks in a product ranged from 2 to 14, the number of products in an order ranged from 1 to 400, the number of resource set is 22, and the average number of tasks in an order is 90.

3.8.3 GA Configuration and Convergence Performance

This subsection discusses methods used to derive the configuration of the GA. We adopted the weights in constructing the next generation for reproduction, mutation, and crossover as 30 %, 20 %, and 50 %, respectively, and a population size of 100 since they were found to provide the best convergence performance.

In the following, we discuss how to determine the relationship between parameters such that an acceptable local optimal solution for each GA run can be achieved within one *TAKT* interval.

Reproduction and mutation are the key processes that increase the probability of convergence to a globally optimal solution [32, 33]. However, it still remains an open question as to what values of GA parameters (such as population size, choice of GA operators, operator weights, and others) to use for a specific problem. Based on the studies in [34], it is recommended that crossover is the key operator that should be used for generating most chromosomes in the next generation.

Since GA is a stochastic optimization process, even over the same order set, GA runs seldom converge to the same solution. We therefore consider results that are averaged over 15 runs. In our production scheduler application, the termination criterion is one TAKT interval, i.e., each GA run lasts for 15 min. We used RPI real-life orders for testing the convergence performance.

Simulation results presented in Fig. 3.6 show that as the mutation ratio decreases, variance of the fitness value also decreases. A high reproduction ratio increases the likelihood of early convergence. A relatively low reproduction ratio decreases the convergence rate.

We next optimize the population size. Studies have shown that satisfactory results can be obtained with a small population size, i.e., the population size does not have to increase exponentially with the complexity of the problem [35]. The complexity of our problem increases exponentially with the number of orders.

Simulation results presented in Fig. 3.7 show that the time to complete one generation increases linearly with the size of the population. The larger the population size, the fewer the generations during one *TAKT* interval. A small population size causes premature convergence even though it has the most number of generations. Fewer generations limit the convergence performance for a larger population size. In addition, a long duration for reach generation should be avoided if new orders arrive frequently since every new order interrupts one dispatching GA thread. Furthermore, no clear dependency between order set size and population size has been found. Figure 3.8 shows that for an order set of 100 orders, GA convergence performance does not clearly depend on the population size.

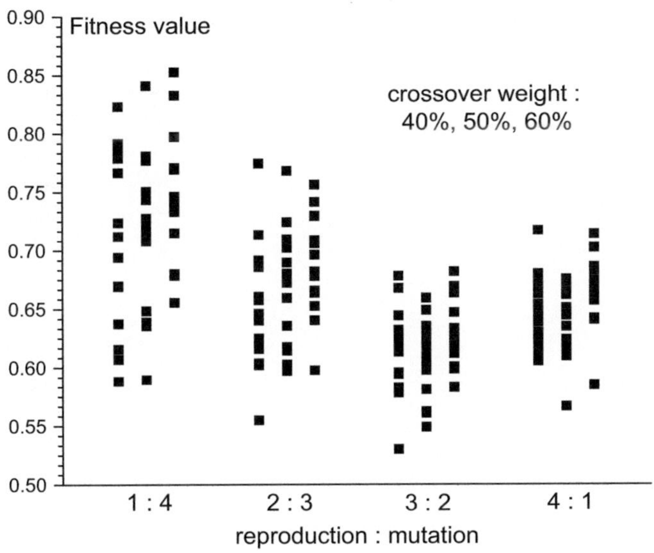

Fig. 3.6 Impact of operator weights on GA convergence performance (termination criterion is one *TAKT* interval)

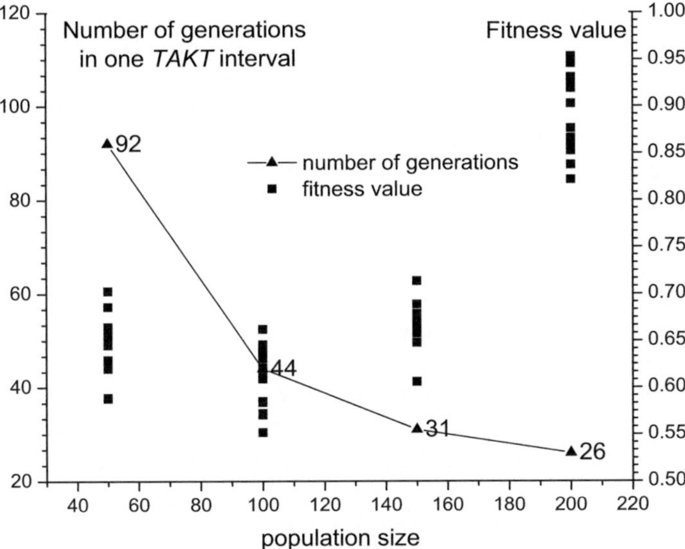

Fig. 3.7 Effect of population size on GA convergence performance (termination criterion is one *TAKT* interval)

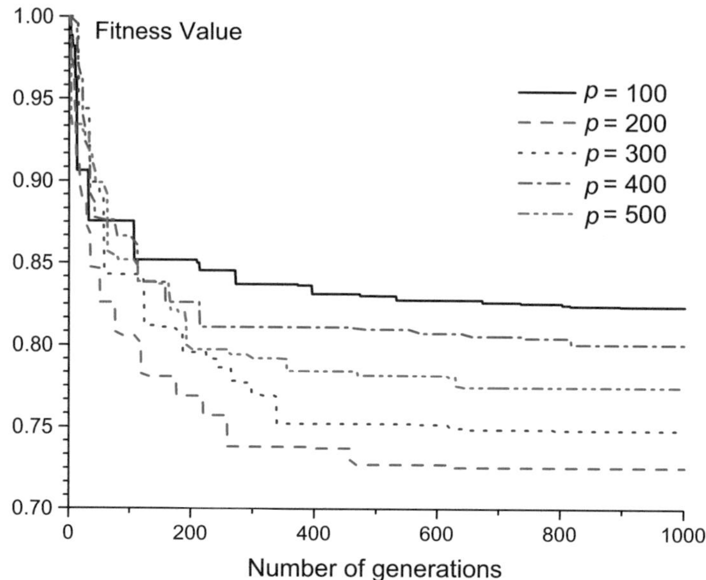

Fig. 3.8 GA convergence performance as a function of population size (p is the population size)

After setting the weights for operators, and the population size, we investigate how the convergence performance varies with order sets of different sizes. Simulation results presented in Fig. 3.9 shows that the time for each generation increases linearly with the number of orders. Therefore, the average number of generations during the same *TAKT* interval decreases with the number of orders, whereas the variance in fitness value increases. Smaller the order set, the lower is the fitness value. This can be attributed to two factors. One reason is that the slack time allocated to each order in a smaller order set is more than it is in a larger order set. The second reason is that more generations are allowed for searching an optimal solution during the same interval.

We next evaluate the computational cost (time per generation) of parallel GAs. Population size is set to 100 regardless of order set size. Simulation results presented in Fig. 3.10 show that, since parallel threads are introduced to handle portions of a big order set, as the number of threads increases, computational cost decreases as well. The decrease rate reaches the highest from a single thread to two parallel threads.

A brute-force method that exhaustively computes the fitness value for each order priority sequence is feasible only when an order set has no more than eight orders. The GA can find the same optimal solution as the brute-force method. For larger problem instances, as shown in Fig. 3.11, GA converges to at least locally optimal solutions after sufficient numbers of generations.

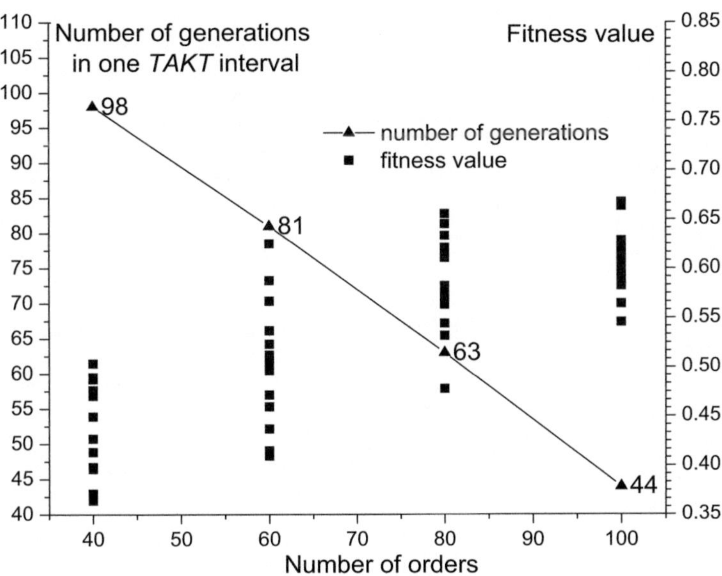

Fig. 3.9 GA convergence performance as a function of orders in an order set (termination criterion is one *TAKT* interval)

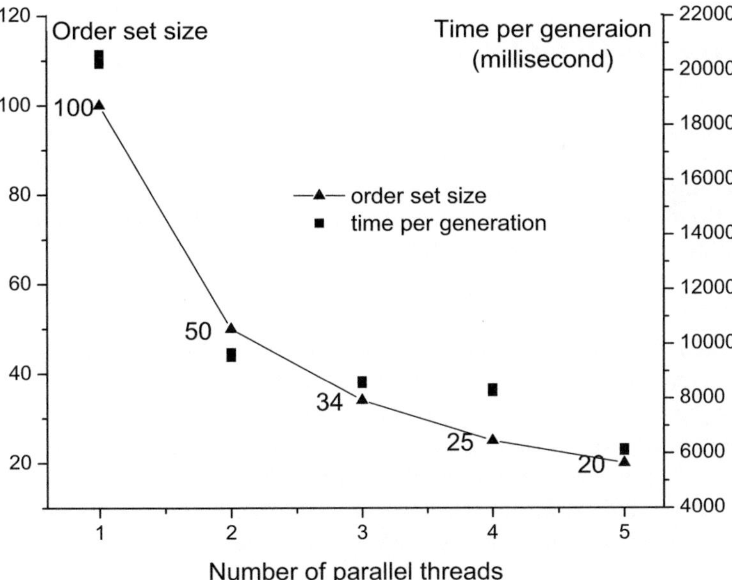

Fig. 3.10 GA convergence performance varies with the number of parallel threads

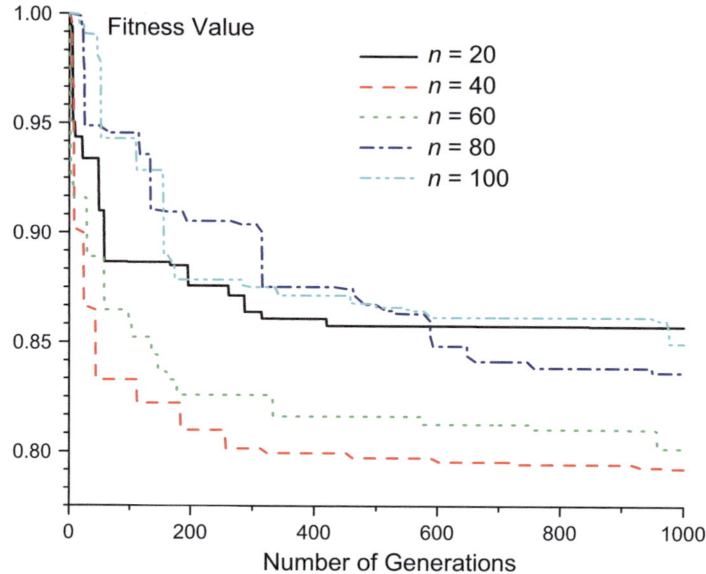

Fig. 3.11 GA convergence performance as a function of generations (n is the number of orders in an order set)

3.8.4   ILP Model for GA Performance Evaluation

In order to evaluate the quality of the results obtained by the GA algorithm in terms of how close it is to globally optimized results, we have developed an integer linear programming (ILP) model that carries out global optimization for the same objective function Eq. (3.11). Notations used in the ILP model are the same as defined for the GA algorithm in Section 3.2. Details of the ILP model are listed below.

1. Constants:

   - Array $PT = [PT_i]$, where i denotes task i and PT_i is the set which contains task $i's$ parent tasks. For the first task of any order, PT_i is empty.
   - Array $ES = [ES_i]$, where ES_i is the earliest possible start time of task i. The value of ES_i for the first task of an order is the manufacturing release time for the order set. For other tasks, the value of ES_i is calculated under the ideal scenario when sufficient resources are available, i.e., there are no WIP tasks and no resource contention for tasks of the same type.

2. Variables:

   Variables in the ILP model are either binary or non-binary variables. They are listed below:

$$s_i : \text{ start time of task } i \text{ (integer variable)}$$

$$e_i : \text{ end time of task } i \text{ (integer variable)}$$

$$r_{ij} = \begin{cases} 1 & \text{if task } i \text{ is assigned to resource } j, \\ 0 & \text{otherwise.} \end{cases}$$

$$\alpha_{ik} = \begin{cases} 1 & \text{if } (s_i \leq s_k), \text{ i.e.,task } i \text{ starts} \\ & \qquad \text{no later than task } k, \\ 0 & \text{otherwise.} \end{cases}$$

$$\beta_{ik} = \begin{cases} 1 & \text{if } (e_i \geq s_k), \text{ i.e., task } i \text{ ends} \\ & \qquad \text{no earlier than task } k, \\ 0 & \text{otherwise.} \end{cases}$$

$$\delta_{ik} = \begin{cases} 1 & \text{if task } i \text{ overlaps with task } k \text{ and} \\ & \qquad \text{task } i \text{ ends no later than task } k \\ 0 & \text{otherwise.} \end{cases}$$

3. Constraints:

Constraints in the ILP model include both linear and nonlinear constraints. Linear constraints are as follows:

$$\sum_{j=1}^{J} r_{ij} = 1, \tag{3.14}$$

$$e_i - sw_i \geq e_k, \ \forall \, k \subset PT_i \tag{3.15}$$

$$e_i - sw_i \geq ES_i. \tag{3.16}$$

Constraint (3.12) ensures that each task has a resource assigned to it. Constraint (3.13) and (3.14) ensure that each task cannot start earlier than its parent tasks or its earliest possible start time. The nonlinear constraints are:

$$\delta_{ik} = \alpha_{ik} \times \beta_{ik} \tag{3.17}$$

$$(s_i - s_k)(1 - \alpha_{ik}) + \alpha_{ik} \geq 1 \tag{3.18}$$

$$\alpha_{ik}(s_i - s_k) \leq 0 \tag{3.19}$$

$$(s_k - e_i)(1 - \beta_{ik}) + \beta_{ik} \geq 1 \tag{3.20}$$

$$\beta_{ik}(s_k - e_i) \leq 0 \tag{3.21}$$

$$\delta_{ik}(1 - r_{ij} - r_{kj}) \geq 0 \tag{3.22}$$

Constraint (3.15) models tasks that overlap with each other. Constraint (3.16) and (3.17) force binary variable α_{ik} to be 1 when task i starts no later than task k. Constraint (3.18) and (3.19) force binary variable β_{ik} to be 1 when task i finishes no earlier than task k. Constraint (3.20) guarantees that under the situation when task i and task k have a time conflict and the possibility of using the same resource, they cannot be assigned to the same resource at the same time.

In order to linearize constraint (3.15), new binary variable γ_{ij} is introduced to replace variable δ_{ij} by standard techniques [36]. The following two constraints are also added to replace constraint (3.15) for all tasks.

$$\alpha_{ij} + \beta_{ij} \leq \gamma_{ij} + 1, \tag{3.23}$$

$$\alpha_{ij} + \beta_{ij} \geq 2 \times \gamma_{ij}. \tag{3.24}$$

Similar techniques are applied to replace product terms involving two variables.

4. Mapping between ILP and GA:

In the ILP model, there is no estimated execution time (introduced in Section 3.2.1). The estimated execution time is calculated as follows:

$$w_i = ES_i - s_i + st_{ij} \tag{3.25}$$

Therefore, in the ILP model the estimated execution time is derived from the estimated earliest start time minus the start time (the subtraction is the estimated wait time), plus the service time. Once the estimated execution time is obtained, the normalized risk-aware slack time can be estimated as described in Section 3.2.

5. Optimality and lower-bound analysis:

The optimal order priority sequence for up to eight orders are the same for the ILP model and the GA algorithm. Due to the large number of constraints, the computation time exceeds 24 h when more orders are considered.

For more than eight orders, we consider ideal and practical lower bounds of the objective function. The ideal lower bound can be obtained under the ideal scenario when sufficient resources are available, i.e., no resource contention. In the ideal scenario, since every order can be scheduled in the desired just-in-time manner, the value of the objective function is zero. The practical lower bound is the best result the GA algorithm can obtain without computation-time limitation. Forty-eight runs of GA were conducted with different combinations of parameter settings. Crossover varies among 40 %, 50 %, and 60 %. The ratio of reproduction and mutation varies among 1:4, 2:3, 3:2, and 4:1. The population size varies among 50, 100, 150, and 200. Instead of being terminated within one *TAKT* interval, GAs were terminated when there was no improvement in the fitness value over 1,000 consecutive generations. The best result of those 48 GAs was referred to as the practical lower bound. Note that, the value of the practical lower bound varies with each problem instance.

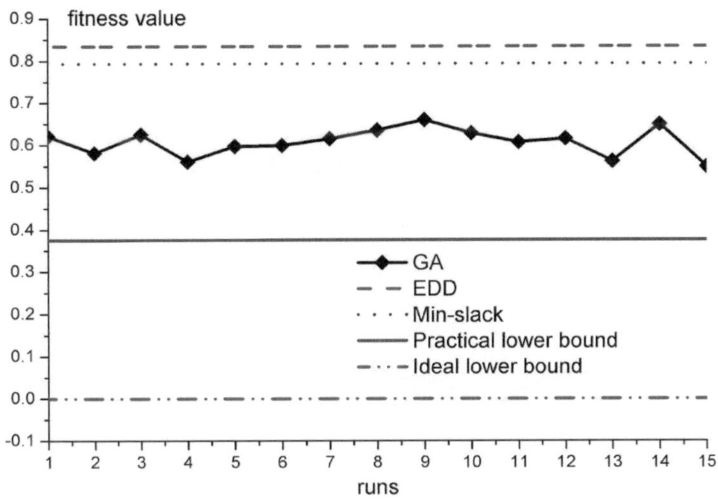

Fig. 3.12 An example of GA performance compared to ideal and practical lower bounds of the objective function

Furthermore, as discussed in Section 3.3.4, two candidate chromosomes that encode EDD and min-slack scheduling solutions are generated in the initial population of the GA algorithm to guarantee that the solution is at least as good as what can be obtained from rule-based heuristics.

As discussed in the beginning of Section 3.4.4, we run the performance test based on the optimal configuration of the GA algorithm. Figure 3.12 shows simulation results obtained from 15 runs for the same order set which contains 100 orders (the maximum number of orders each GA thread is configured to handle).

Finding optimal solutions to our problem is not practical due to the problem complexity, e.g., the ILP model for large problem instances takes vast amount of computation time. The ILP model for small problem instances shows that the GA algorithm can obtain optimal results for small problem instances. The lower-bound performance analysis for large problem instances shows that although the GA algorithm cannot obtain the result as good as the ideal lower bound, its performance is superior to rule-based heuristics. In addition, if more computation time is allowed, GA can obtain even better performance.

3.8.5   Production Scheduler Configuration

Multiple GA threads are run in parallel, which contains one scheduling GA and the others are dispatching GAs. Figure 3.13 presents results for two parallel dispatching GA threads. Comparison results for different number of dispatching GAs are presented in Table 3.1b. All the evaluation metrics are evaluated over an interval

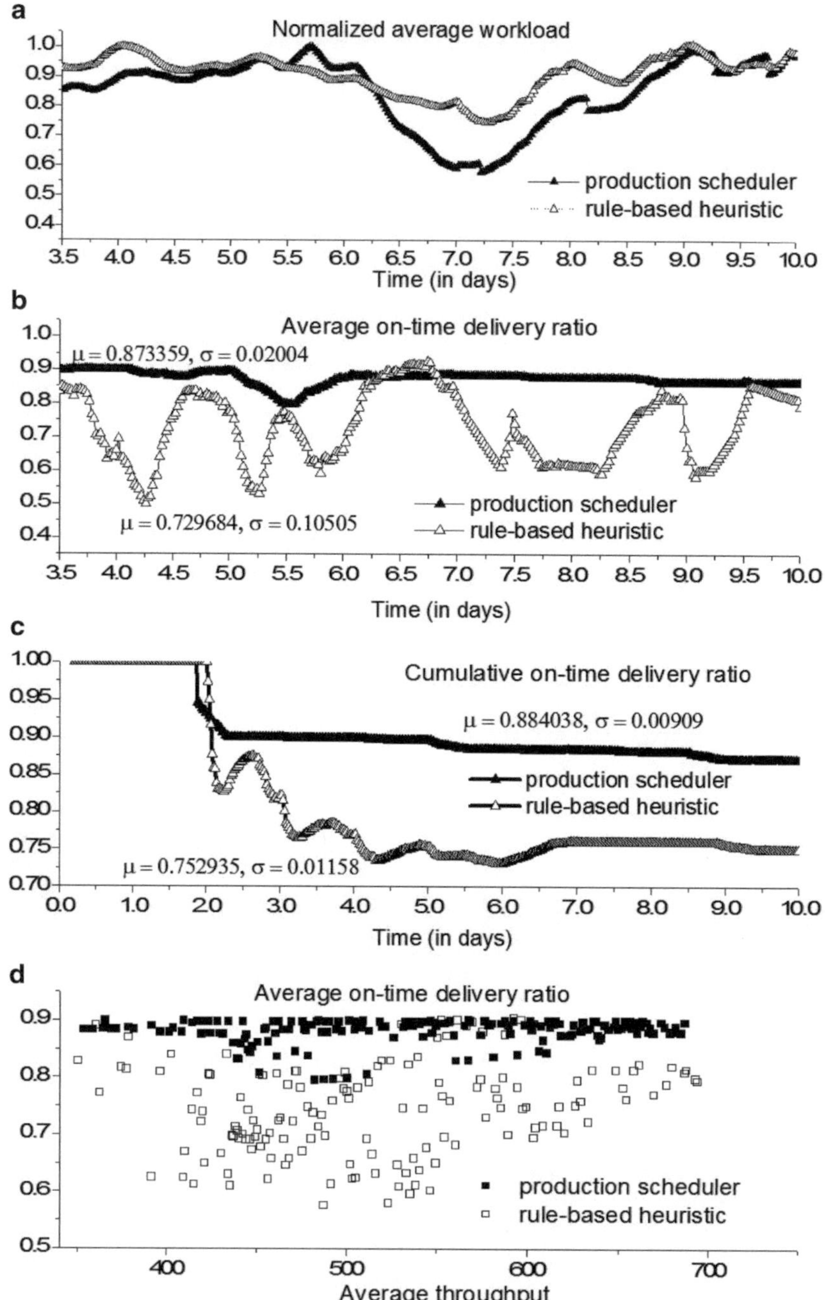

Fig. 3.13 VPF simulation results for a single run using real-life order data

Table 3.1 Mean (with standard deviation in parentheses) of average and cumulative on-time delivery ratio for (a) different simulation runs using real-life and synthetic order data; (b) varying number of dispatching GA threads for the production scheduler

(a)

Algorithm	Performance metrics on-time delivery ratio	Run 1	Run 2	Run 3	Run 4	Run 5
Production	Average	0.87336	0.89095	0.89457	0.88276	0.88806
		(0.02004)	(0.00650)	(0.00623)	(0.01322)	(0.00977)
Scheduler	Cumulative	0.88404	0.89037	0.89002	0.88715	0.88864
		(0.00909)	(0.00654)	(0.00790)	(0.00802)	(0.00774)
Min-slack	Average	0.72968	0.74717	0.74715	0.73842	0.74279
		(0.10505)	(0.11592)	(0.11594)	(0.11048)	(0.11321)
Heuristic	Cumulative	0.75274	0.74986	0.75033	0.75138	0.75078
		(0.01158)	(0.01460)	(0.01778)	(0.01362)	(0.01517)

(b)

Performance metrics	1 thread	2 threads	3 threads	4 threads	5 threads
Average on-time delivery ratio	0.87071	0.87336	0.86477	0.77777	0.75154
Average on-time delivery ratio	(0.02745)	(0.02004)	(0.02885)	(0.06933)	(0.09845)
Cumulative on-time delivery ratio	0.88176	0.88404	0.87506	0.78264	0.76087
Cumulative on-time delivery ratio	(0.01533)	(0.00909)	(0.01279)	(0.00778)	(0.01226)

of 6 h. Intervals ranging from 1 to 12 h show similar results. A single-run simulation ranges from 7 to 10 days in terms of factory time. Order due dates are tight. For loose due dates, any scheduling method can be used.

3.8.6 Results and Discussions

1. Results from a single run of real-life RPI data

We consider factory optimization for a real-life order pool provided by RPI. The performance of the production scheduler and the rule-based heuristic are compared. Figure 3.13a shows the normalized average workload. Figure 3.13b shows the average on-time delivery ratio. Figure 3.13c shows cumulative on-time delivery ratio, where the interval is from the start of factory operation to the current time. Figure 3.13d shows the average on-time delivery ratio versus average throughput. The results are listed in Table 3.1a ("Run 1").

Based on the real-life order data from RPI, the following conclusions can be drawn from simulation results:

(i)   Figure 3.13a shows that factory workloads resulting from the two scheduling algorithms are different. This is because of the difference in order dispatching mechanisms of the two scheduling algorithms. Rule-based heuristics instantly generate a schedule for an order, which gives the exact time when this order should be dispatched for manufacturing. The production scheduler however, does not dispatch an order right after its arrival. As discussed in Sect. 3.3, it takes at least two TAKT intervals before an order can be dispatched. Figure 3.13b shows that the proposed schedule provides a higher on-time delivery ratio than the rule-based heuristic used by RPI. Moreover, it provides an on-time delivery ratio with less variance. In contrast, the performance of the rule-based heuristic is highly variable (and hence unpredictable).

(ii) Figure 3.13b–c show that the production scheduler can stabilize and improve the on-time delivery ratio significantly compared to rule-based heuristics. Furthermore, Fig. 3.13c shows that the cumulative on-time delivery ratio of the production scheduler is always higher and more stable than that for rule-based heuristics. This is because the production scheduler is a global scheduler. It generates a schedule for an order while considering the influence of scheduling other orders as well as current factory status; important considerations that are ignored by the rule-based heuristic currently deployed in practice.

(iii) Figure 3.13a–b show that, as expected, an increase (decrease) in workload always leads to a decrease (increase) in on-time delivery ratio regardless of scheduling algorithms. This is because a heavier workload of the factory increases the amount of tasks accumulated in resource queues, which subsequently increases the average execution time of an order. The likelihood of an order missing its due date is proportional to the factory workload during the period when it was being processed. Note however that the production scheduler is able to handle high workloads more efficiently.

(iv) Figure 3.13d shows that for the same factory throughput, the on-time delivery ratio resulting from the production scheduler is always larger than or comparable to that for the rule-based heuristics. The variation of on-time delivery ratio of the production scheduler is also less. When the same number of orders have been delivered during the same interval, the production scheduler results in fewer late orders in spite of the computation time required by GAs. These results demonstrate the higher scheduling efficiency of the production scheduler.

2. Results from other runs

We ran the VPF using the same RPI data two additional times. Compared to the first run, the amount of rework is different because of the stochastic nature of fault occurrences. Simulation results for the second and third run of the real-life RPI data are included in Table 3.1a.

We deployed the production scheduler for the fourth and fifth run on data sets that have been generated based on the real-life RPI data. We randomly select pairs of orders and randomly exchange some attributes except for due dates. Simulation results are also shown in Table 3.1a ("Run 4" and "Run 5").

3. Parallel performance

Simulation results for parallel dispatching GAs on the RPI order data are presented in Table 3.1b. The following conclusions can be drawn from simulation results:

(i) Two parallel dispatching GA threads are more effective then a single thread, as well as more than two threads. Two parallel threads leads to higher on-time delivery ratio and less variance in the results. As the number of orders to be processed during one *TAKT* interval increases, the probability that a single thread cannot converge to an effective solution increases. Two parallel threads partition the order set, therefore, each of them has a higher probability of converging to an optimal solution over the half of the original problem. However, note that orders are partitioned according to order due date. When there are more than two threads, the assembled solution from each thread can no longer capture other factors that influence order priorities.

(ii) The optimal number of dispatching GAs depends on problem attributes such as order pool size, due date, and order attribute diversity. The book-keeping and merging overhead resulting from an unnecessarily large number of threads offsets the benefit gained from parallelism. In terms of RPI's dataset, two parallel dispatching GA threads is the most appropriate setting for the production scheduler. A theoretical characterization of the optimum number of threads in general cases needs more study and it is left for future work.

3.9 Conclusion

We have presented a production scheduler based on an IGA. The production scheduler is used to provide an effective dispatching sequence of orders and products, and simultaneously assign resources to tasks. It adopts an distributed computing framework using parallel threads. Each thread provides a partial solution and the global solution is integrated from partial solutions. The production scheduler also dynamically ingests the up-to-the-moment factory information to update parameters used for computation. We have implemented the production scheduler in a virtual print factory, and compared it to a heuristic method that is currently deployed in practice. We have shown that the production scheduler clearly outperforms the rule-based heuristic for a real-life order set and synthetic order set.

In future work, we will further investigate the stochastic nature of the optimization problem. We will study machine-learning techniques to identify outliers, tolerate noise in the factory and order data, and provide predictions with probabilities that can be incorporated in decision-making for scheduling and resource allocation.

References

1. J. Zeng, S. Jackson, I. Lin, M. Gustafson, E. Gustafson, R. Mitchell, Operations simulation of on-demand digital print, in *IEEE 13th International Conference on Computer Science and Information Technology*, Chengdu (2011)
2. S.P. Hoover, G.A. Gibson, The future of print and the digital printing revolution, in *31st International Conference on Imaging Science*, Beijing (2010)
3. J.K. Lenstra, A.H.G.R. Kan, P. Brucker, Complexity of scheduling under precedence constraints. Oper. Res. **26**, 22–35 (1978)
4. J. Zeng, I.-J. Lin, E. Hoarau, G. Dispoto, Productivity analysis of print service providers. J. Imaging Sci. Technol. **54**(6), 1–9 (2010)
5. J. Zeng, I.-J. Lin, E. Hoarau, G. Dispoto, Next-generation commercial print infrastructure: Gutenberg-Landa TCP/IP as cyber-physical system. J. Imaging Sci. Technol. **54**(1), 1–6 (2010)
6. M.P. Fanti, B. Maione, G. Piscitelli, B. Turchiano, System approach to design generic software for real-time control of fexible manufacturing systems. IEEE Trans. Syst. Man Cybern. Part A: Syst. Hum. **26**(2), 190–202 (1996)
7. M. Dotoli, M.P. Fanti, C. Meloni, M. Zhou, Design and optimization of integrated e-supply chain for agile and environmentally conscious manufacturing. IEEE Trans. Syst. Man Cybern. Part A: Syst. Hum. **36**(1), 62–75 (2006)
8. C. Özgven, L. Özbakir, Y. Yavuz, Mathematical models for job-shop scheduling problems with routing and process flexibility. Appl. Math. Model. **34**, 1539–1548 (2010)
9. J. Zeng, I.-J. Lin, G. Dispoto, E. Hoarau, G. Beretta, On-demand digital print services: a new commercial print paradigm as an it service vertical, in *Annual SRII Global Conference*, Silicon Valley (2011), pp. 120–125
10. M. Agrawal, K. Chakrabarty, J. Zeng, I.-J. Lin, G. Dispoto, Simultaneous task scheduling and resource binding for digital print automation, in *Institute of Industrial Engineers (IIE) Industrial Engineering Research Conference*, Reno (2011)
11. K. Trivedi, *Probability and Statistics with Reliability, Queuing and Computer Science Application*, 2nd edn. (Wiley, New York, 2001)

12. G. Grimmett, D. Stirzaker, *Probability and Random Processes* (Oxford University Press, Nairobi, 2001)
13. S. Tzafestas, A. Triantafyllakis, Deterministic scheduling in computing and manufacturing systems: a survey of models and algorithms. Math. Comput. Simul. **35**, 397–437 (1993)
14. L. Shi, S. Olafsson, Nested partitioned method for global optimization. Oper. Res. **48**, 390–407 (2000)
15. H. Fei, N. Meskensa, C. Chu, A planning and scheduling problem for an operating theatre using an open scheduling strategy. Comput. Ind. Eng. **58**(2), 221–230 (2012)
16. P.M. Sanderson, The human planning and scheduling role in advanced manufacturing systems: an emerging human factors domain. Hum. Factors **31**, 635–666 (1989)
17. X. Cai, S. Zhou, Sequencing jobs with random processing times to minimize weighted completion time variance. Ann. Oper. Res. **70**, 241–260 (1997)
18. K.R. Baker, G.D. Scudder, Sequencing with earliness and tardiness penalties: a review. Oper. Res. **38**, 22–36 (1990)
19. J. Blazewicz, J.K. Lenstra, A.H.G.R. Kan, Scheduling subject to resource constraints: classification and complexity. Discret. Appl. Math. **5**(1), 11–24 (1983)
20. T. Hegazy, Optimization of resource allocation and leveling using genetic algorithms. J. Constr. Eng. Manage. **125**(1), 167–175 (1999)
21. N. Mansour, M. Awad, K. El-Fakih, Incremental genetic algorithm. Int. Arab J. Inf. Technol. **3**(1), 42–47 (2006)
22. M. Srinivas, L.M. Patnaik, Genetic algorithms: a survey. IEEE Trans. Comput. 27, 17–27 (1994)
23. T. Murata, H. Ishibuchi, H. Tanaka, Multi-objective genetic algorithm and its applications to flowshop scheduling. Comput. Ind. Eng. **30**, 957–968 (1996)
24. H.C.W. Lau, T.M. Chan, W.T. Tsui, W.K. Pang, Application of genetic algorithms to solve the multidepot vehicle routing problem. IEEE Trans. Autom. Sci. Eng. **7**(2), 383–392 (2009)
25. Z. Yang, P.I. Rockett, Application of multiobjective genetic programming to the design of robot failure recognition systems. IEEE Trans. Autom. Sci. Eng. **6**(2), 372–376 (2008)
26. O. Sobeyko, L. Mönch, Genetic algorithms to solve a single machine multiple orders per job scheduling problem, in *Proceedings of the 2010 Winter Simulation Conference*, Baltimore (2010), pp. 2493–2503
27. J.C. Bean, Genetics and random keys for sequencing and optimization. ORSA J. Comput. **6**(2), 154–160 (1994)
28. W.M. Spears, K.A. Dejong, *On the Virtues of Parameterized Uniform Crossover* (Naval Research Lab, Washington, DC, 1995)
29. C.-Y. Lee, G. Vairaktarakis, Minimizing makespan in hybrid flowshops. Oper. Res. Lett. **16**(3), 149–158 (1994)
30. J. Brenner, G. Schäfer, Cost sharing methods for makespan and completion time scheduling, in *Proceedings of the 24th International Symposium on Theoretical Aspects of Computer Science (STACS)*, Aachen. Lecture Notes in Computer Science, vol. 4393, no. 3 (2007) pp. 670–681
31. S. Mito, T. Ohno, *Just-In-Time for Today and Tomorrow* (Productivity Press, Cambridge, 1986)
32. G. Rudolph, Convergence analysis of canonical genetic algorithms. IEEE Trans. Neural Netw. **5**(1), 96–101 (1994)
33. D. Bhandari, C.A. Murthy, S.K. Pal, Genetic algorithm with elitist model and its convergence. Int. J. Pattern Recognit. Artif. Intell. **10**(6), 731–747 (1996)
34. K. Deb, S. Agrawal, Understanding interactions among genetic algorithm parameters, in *Foundations of Genetic Algorithms*, ed. by W. Banzhaf, C.R. Reeves (Morgan Kaufmann, San Francisco, 1999), pp. 265–286
35. D.E. Goldberg, K. Deb, J.H. Clark, Genetic algorithms, noise, and the sizing of populations. Complex Syst. **6**, 333–362 (1992)
36. H.P. Williams, *Model Building in Mathematical Programming*, 2nd edn. (Wiley, New York, 1985)

Chapter 4
Predictions of Process-Execution Time and Process-Execution Status

As shown in Chap. 3, process-execution time is a fundamental measure in an EIS. Our risk-aware execution-time estimation method (Sect. 3.2.2) has demonstrated improved performance over static rule-based methods. However, in addition to performing real-time production scheduling, an EIS should also be able to carry out planning for the future. Therefore, accurate predictions of both process-execution time and process status are crucial for the development of an intelligent EIS. We propose new process-execution time-prediction and process status-prediction methods for an EIS. Process-execution time prediction is a regression problem and state-of-the-art (baseline) time-prediction methods use a machine-learning regression model. Process status prediction is a binary classification problem in which a class labeled "completed" or "in-progress" is assigned to a process with respect to an arbitrary predictive horizon (i.e., the future time given by the method user). A state-of-the-art (baseline) status-prediction method compares the predicted process-execution time with the predictive horizon to check whether the process is still in progress. Another baseline method builds individual status-prediction classifiers for all possible predictive horizons. The methods proposed in this chapter are developed on the basis of the baseline time-prediction methods by integrating statistical methods with machine-learning algorithms. Comparison results obtained from the real data of a digital-print enterprise show that the proposed time-prediction method reduces both the relative mean error and the root-mean-squared error of the regression model. Furthermore, the proposed status-prediction method not only achieves higher classification accuracy than both state-of-the-art methods, it also estimates the probability of the predicted status. In addition, algorithm development and training phases of the proposed methods do not rely on any arbitrary predictive horizon. Therefore, a single time-prediction model as proposed is sufficient for status prediction as opposed to a baseline status-prediction method that requires classification models for all potential predictive horizons.

© Springer International Publishing Switzerland 2015

Q. Duan et al., *Data-Driven Optimization and Knowledge Discovery for an Enterprise Information System*, DOI 10.1007/978-3-319-18738-9_4

The rest of this chapter is organized as follows. Section 4.1 discussed the background and motivation. Section 4.2 introduces the proposed prediction framework, problem statement, and the real data source used in this chapter. Section 4.3 describes in detail the state-of-the-art, the proposed time-prediction method, and the performance of method. Section 4.4 describes in detail the state-of-the-art, the proposed status-prediction method, and the performance of method. Finally, Sect. 4.5 presents conclusions and outlines ideas for further work. Since SVR is the key solution technique used in this chapter, an introduction to it is provided in Appendix D.

4.1 Introduction

An enterprise offers diverse, and often on-demand, customized services. Each service can involve multiple processes, which may be parallel, in series, or in any combination thereof. A process is defined as any enterprise operation that requires a resource. Process-execution time is defined as the time spent by the enterprise executing that process. The status of a process can be either "completed" or "in-progress" with respect to an arbitrary predictive horizon (i.e., the future time of interest) [1].

An enterprise information system (EIS) is a real-time embedded system within an enterprise and is responsible for recording, monitoring, and predicting the attributes related to each process. For system-wide design and optimization, time and process status serves as the basis for a number of performance measures [2]. These measures are very informative for making decisions or planning schedules. For example, Fig. 4.1 shows the multiple processes (printing, trimming, sewing, binding, etc.) that are involved in producing two products in a digital print factory. Delivering a large amount of different products on time requires the EIS to perform accurate process-time predictions and further use these predictions to perform scheduling and resource allocation [3]. The potential relevance of digital print optimization to electronic design automation was recently highlighted in [4]. Additional examples in electronic design automation include: (i) performance optimization in the design of 3D integrated circuits requires through-silicon via (TSV) delay prediction [5]; (ii) network- on-chip system design requires channel and switch congestion time and status prediction [6]; (iii) design-space exploration requires the application time and fitness status prediction [7]; (iv) communication system design and optimization requires the prediction of message response time [8].

With precise time and status prediction, an EIS can suggest optimal resource allocation for processes and estimate resource availability, allowing managers to make higher-level management decisions based on predicted process-execution times for lower-level processes [9]. A detailed scheduling and resource allocation algorithm using estimates of process-execution times is presented in [3]. Process time prediction can be used to schedule many other activities, such as maintenance interruptions. Being able to predict the time and status of a process in real time is

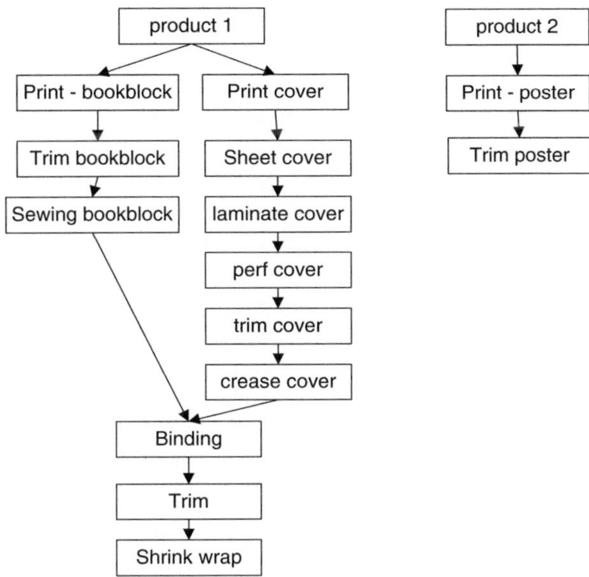

Fig. 4.1 Illustration of multiple processes involved in producing two products in a digital print enterprise

critical for service and process-flow management in an EIS [10]. Traditionally, it is sufficient to make time and status predictions relying on pre-designed rule-based specification or even human judgment. However, increased system complexity and heterogeneity necessitate advanced algorithms and real-time performance prediction for embedded systems [11–13].

If we take the digital print industry as an example, a use case can be as follows. During the morning shift, the floor manager looks at the queue of a web-press. Suppose there are 14 jobs in the queue. The manager might want to know how many of them will be left for the afternoon shift. Another use case can be a process-status dashboard that periodically predicts which processes in the queue can be completed within the next hour [14].

Process-execution time and process status are influenced by many features. For instance, a resource malfunction event delays scheduled processes. An older-version resource may take more time to execute a process than a newer version of the resource. Process workload also affects execution time. Related processes, e.g., processes that share the same resources, may influence the execution times of one another. These dynamic features make time and status predictions very complex, making it difficult to achieve high accuracy [15]. Furthermore, features for making predictions are selected through several iterations of data selection, pre-processing, model building, and model analysis.

Time prediction has been studied in many EISs. A common traditional approach is based on the multiple-factor linear combination (MFLC) method [16]. This solution is extremely inaccurate when the attributes of a process are dynamic, e.g., when manual work is involved in a process. It is obvious that it cannot be adopted in transient, heterogeneous, and stochastic enterprise environments. A statistical modeling method that considers uncertain factors and nonlinear relationships between features is a significant improvement over the MFLC method [16–18]. Although this solution does not require a clear understanding of the underlying physical processes, assumptions made for building the model must be able to depict the actual situation as clearly as possible.

Production simulation is another advanced solution for process-execution time and process status predictions [19]. However, this solution requires a clear understanding of the underlying physical processes. Developing models for simulation can be expensive and the execution can be lengthy if the predictive horizon is long [20]. Therefore, production simulation for execution-time prediction is not sufficiently general or flexible for most EISs [21].

Recently, data-driven methods have become mainstream [22]. An advantage of data-driven approaches is that they do not require extensive expertise in process modeling. These data-driven methods correlate historical, current, and future data, without explicitly addressing the physical processes in the way model-based approaches do. There are a number of examples in the literature of successful application of time and status predictions; for instance, network-on-ship latency evaluation based on support vector regression (SVR) [23] and software execution time prediction and performance evaluation based on regression models [24], aircraft taxi time prediction based on different machine-learning algorithms [25], and tumor status prediction based on SVR [26].

Recent work has therefore shown that baseline process-execution time prediction can be realized by using a machine-learning regression model. However, the prediction performance fluctuates and even deteriorates as predictions become more complex. Errors accumulate for predictions made using other predicted values as input information. We consider two start-of-the-art status-prediction methods as baselines in this work. The first baseline status-prediction method compares the predicted process-execution time with the predictive horizon to check whether the process is still in progress. A drawback is that it does not differentiate potential impact of the predictive horizon. Intuitively, the larger the predictive horizon, the higher the probability it is that a process can be completed by that predictive horizon. The second baseline status-prediction method builds individual status-prediction classifiers for all possible predictive horizons. The disadvantage with this approach is that it needs to know the predictive horizons in advance in order to train each classifier [15].

In this chapter, we design a time-prediction method and a status-prediction method. The proposed methods provide higher prediction accuracy with only one regression model.

The main contributions of this chapter are as follows.

- By integrating statistical analysis with machine-learning algorithms, the proposed time-prediction method takes into consideration the performance deviation of the baseline time-prediction method with respect to different instances. It modulates the initial predicted value to get the final predicted value such that

higher prediction accuracy can be achieved. The proposed method also absorbs the effect of outliers as long as a sufficient number of testing instances is provided.
- Along with a binary process status prediction, the proposed status-prediction method also estimates the probability of the predicted status. It further enables conditional prediction if the intermediate process status is considered to be new evidence.
- By replacing multiple classification models with a single regression model, the proposed status-prediction method greatly reduces the cost of building extra classification models. It also considers the potential impact of the arbitrary nature of a predictive horizon.

4.2   Problem Statement and Data Source

4.2.1   Problem Statement

Figure 4.2 is an illustration of the proposed process-execution time and process status prediction framework. The framework can be established for any type of process whose execution time and status information is important for production management. Given an arbitrary predictive horizon, the framework is able to predict in real time the status and the probabilities of the status for the processes in the resource queue.

4.2.1.1   Time-Prediction Problem Statement

Process-execution time prediction is formulated as a regression problem. Each data instance is denoted as $(\mathbf{x}, y)$, where vector $\mathbf{x}$ is the feature vector of the process and the numerical value y is the prediction target, i.e., the execution time of the process. For instance, the number of pages, the width of one page, and the height of one page can be three features of a printing process. Given a feature vector $\mathbf{x} = (200, 11, 17)$, the proposed method may predict, for example, that it takes $214\,\mathrm{s}$ to print 200

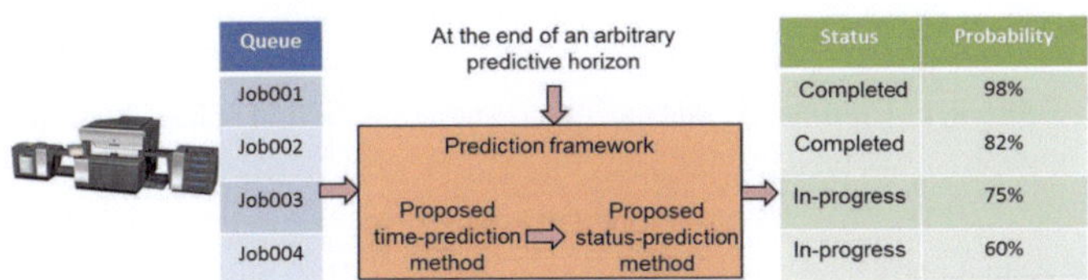

Fig. 4.2 Illustration of the proposed process-execution time and process status prediction framework

A4 pages. The feature selection and construction process of the feature vector $\mathbf{x}$ are explained in detail in Sect. 4.2.4. A regression analysis process, which estimates the relationships among input features, has to be conducted in order to predict the numerical value of the process-execution time.

Relative mean error (RME) and root-mean-squared error (RMSE) are performance indices that are commonly used to evaluate a regression model [27]. We took at least 4,000 samples in all the following tests in order to guarantee that RME and RMSE demonstrate the actual performance. They are defined in Eqs. (4.1) and (4.2), respectively:

$$RME = \frac{1}{N} \sum_{i=1}^{N} \left| \frac{y_i - y_i^*}{y_i} \right| \tag{4.1}$$

$$RMSE = \sqrt{\frac{1}{N} \sum_{i=1}^{N} \left| \frac{y_i - y_i^*}{y_i} \right|^2} \tag{4.2}$$

where y_i is the actual value, y_i^* is the predicted value, and N is the number of instances that have been evaluated. The less the error, the higher the model accuracy.

4.2.2 Status-Prediction Problem Statement

Process status prediction is formulated as a classification problem. Each data instance is denoted as $(\mathbf{x}, y)$, where vector $\mathbf{x}$ is the feature vector of the process as defined in Sect. 4.2.1.1 and categorical value y is the prediction target, i.e., the binary status ("completed" or "in-progress") of the process. For instance, the number of books and the pages of one book can be two features of a binding process. Given a feature vector $\mathbf{x} = (12, 200)$, the proposed method predicts that after 5 min the binding process of 12 200-page books will still be "in-progress" with a probability of 0.9, and after 10 min the binding process will be "completed" with a probability of 0.98.

Table 4.1 shows the confusion matrix which depicts how binary-class predictions on instances can be distinguished. The performance of a classifier can be evaluated by *error rate* which is defined in Eq. (4.3).

Table 4.1 Confusion matrix of a status-prediction classification model

	Predicted class	
Actual class	Positive	Negative
Positive (completed)	True positive (*TP*)	False negative (*FN*)
Negative (in-progress)	False positive (*FP*)	True negative (*TN*)

Fig. 4.3 Event log structure

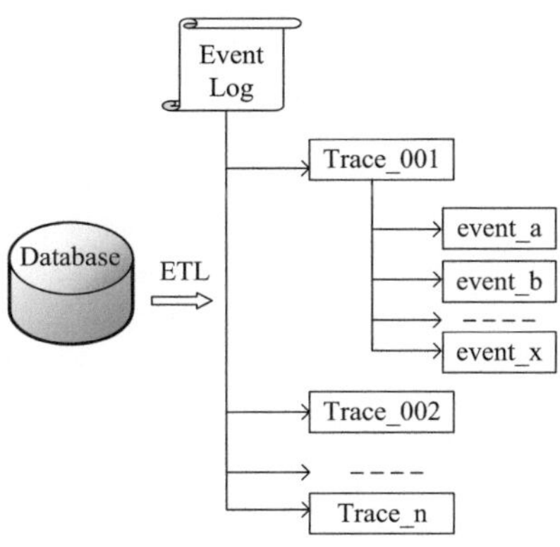

$$error\ rate = \frac{FP + FN}{FP + TP + FN + TN} \tag{4.3}$$

$$error\ percentage = error\ rate \times 100\,\% \tag{4.4}$$

where FP, FN, TP, and TN are described in Table 4.1.

4.2.3   Production Event Log

The surveillance system of an EIS records all the major activities and stores raw information into databases [28]. Before using this raw data, standard database preprocessing needs to be performed, such as extracting useful and correct data, transforming data into an appropriate format, and loading data into the target data warehouse [29]. After preprocessing all the records, we can get an event log organized on the basis of trace and event.

Figure 4.3 shows the event log structure. Each trace corresponds to a service. Each event corresponds to a process. Each event records key information, such as service ID, process ID, process description, the ID of the worker who executed this process and timestamps of the start and end of the process. The precedence relations among processes are reflected by the timestamps associated with each event.

4.2.4   Data Source

In this chapter, we use real data from the EIS production event log of a large multi-national mass customization manufacturer as a test case to demonstrate the building, optimization, and evaluation procedures of the proposed prediction framework.

Table 4.2 Features of the input vector

Feature	Property	Explanation
A_1	Numerical	Number of pages
A_2	Numerical	Width of one page
A_3	Numerical	Height of one page
A_4	Binary	Indicates whether print in color
A_5	Binary	Indicates whether print in black and white
A_6	Binary	Indicates whether print on book paper
A_7	Binary	Indicates whether print on glossy paper
A_8	Binary	Indicates whether print on one side
A_9	Binary	Indicates whether print on both sides
A_{10}	Numerical	Elapsed time since last maintenance
A_{11}	Numerical	Number of executed processes since last maintenance
A_{12}	Numerical	Number of processes in the queue that are ahead of this process
A_{13}	Numerical	Sum of the predicted execution times for processes in the queue that are ahead of this process

The process complexity is correlated to the prediction models. We considered as many inputs as possible to address process complexity. In the feature-selection phrase, we retained only predictive inputs based on our assessment of their importance. Therefore, the more complex the process, the more inputs there are to the prediction model.

In the test case, we have identified 13 ad-hoc features that contribute in predicting the process-execution time and status of one printing process which uses press as the resource. The resulting feature vector $\mathbf{x}=(A_1, A_2, \ldots, A_{13})$ is explained in Table 4.2. For example, feature A_1 describes the number of pages that are needed to be printed in this process, feature A_2 and feature A_3 describe the size of each page, and feature A_{10} describes the time since last maintenance of the press that will be used to perform this process. Features consist of attributes of the process (A_1 to A_9), properties of the resource (A_{10} to A_{11}), and the attributes of the resource queue (A_{12} to A_{13}). This set of features include the attributes of the process (A_1 to A_9) as well as the interactive influence from other processes (A_{10} to A_{13}). Feature data were extracted from EIS event logs and transformed into the right format for building the predictive models.

As shown in Fig. 4.4, the training, testing, and evaluation data were extracted from three consecutive weeks of real data. We used 4,896 instances as the training data to build the model; 4,637 instances as the testing data to tune parameters and optimize the framework performance; and 4,725 instances as the evaluation data to evaluate the performance of our proposed methods.

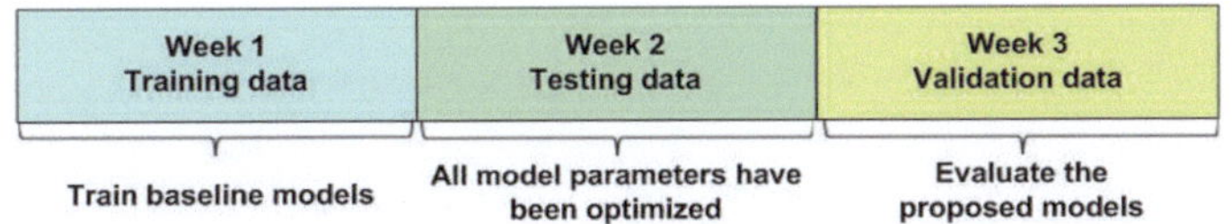

Fig. 4.4 Partitioning of data for training, testing, and validation

4.3   Process-Execution Time Prediction

4.3.1   Baseline Time-Prediction Method

SVR has been shown to provide superior performance in many related research problems [30, 31]. It has also been shown to have better performance than other non-linear learning algorithms, including decision trees and neural networks in our related research studies [32]. An SVR model is adopted as the baseline time-prediction model. An introduction of SVR is presented in Appendix D. We select the popular sequential minimal optimization (SMO) algorithm to train the SVR model since it is known for its fast convergence [33]. Weka, the open source software [34], has been used to realize the algorithm. Parameters of the baseline time-prediction model built from the training data have been tuned such that both RME and RMSE are minimized for the testing data. The baseline time-prediction model has maximum generalization ability for the training data and minimum error for the testing data. Figure 4.5a shows the procedure of the baseline time-prediction method for making a new prediction. The initial predicted time given by the baseline time-prediction model is treated as the result of the baseline method.

4.3.2   Proposed Time-Prediction Method: Integration Based on Statistical Analysis and Machine Learning

The proposed time-prediction method as shown in Fig. 4.5b is designed upon and developed from the baseline time-prediction method. The procedure and the main components of the proposed time-prediction method are discussed as follows.

4.3.2.1   Probability Mass Function (PMF) of the Relative Error

After optimizing the parameters of the baseline time-prediction model, we derive the histogram and the probability mass function (PMF) of the model's relative error. The PMF can be derived from the histogram; probability density is derived by dividing the frequency value by the number of instances. The relative error (RE) for an instance is defined as follows:

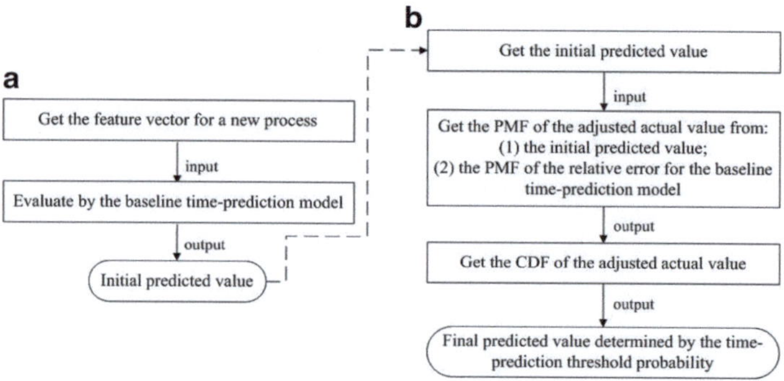

Fig. 4.5 Illustration of (**a**) the baseline time-prediction method and (**b**) the proposed time-prediction method

Fig. 4.6 The histogram and the probability mass function (PMF) of the relative error for the baseline time-prediction model

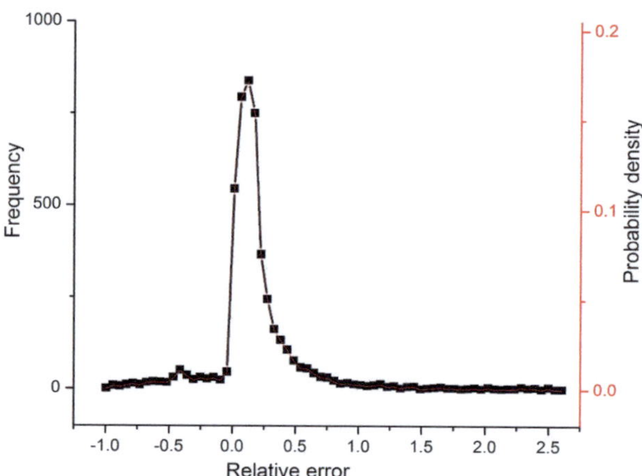

$$\mathrm{RE} = \frac{y^* - y}{y} \tag{4.5}$$

where y is the actual value and y^* is the predicted value given by the baseline time-prediction model. The histogram/PMF of the relative error by evaluating the testing data through the baseline time-prediction model is shown in Fig. 4.6.

The key feature of the baseline time-prediction model is that its histogram or PMF of the relative error is uniquely determined by its performance over the testing data. Therefore, we can see that the histogram/PMF of the relative error can be used to describe the performance of the baseline time-prediction model. For instance, as shown in Fig. 4.6, the distribution of the relative error is highly skewed. The baseline time-prediction model makes overestimation more frequently than underestimation. The overestimation errors are also more severe than underestimation errors.

The more the number of testing instances, the more accurate is the histogram/PMF. Rather than fitting the PMF with any parametrized distribution model, we record the PMF of the baseline time-prediction model. The following procedure is designed to make use of the information provided by the histogram/PMF.

4.3.2.2   Probability Mass Function (PMF) of the Adjusted Actual Value

For every new instance, the baseline time-prediction model generates the initial predicted value. We made an assumption that the relative error of the initial predicted value follows the same distribution of the relative errors derived from historical data maintained by the baseline time-prediction model, i.e., the histogram/PMF of the relative error derived from the testing data. It is a reasonable assumption since the new instance is randomly generated from the same distributions as the testing data.

By transforming Eq. (4.5), we can derive the following relationship:

$$\mathrm{RE} = \frac{y^* - y}{y} \Rightarrow y_a = \frac{y^*}{1 + \mathrm{RE}} \tag{4.6}$$

where y_a is the adjusted actual value of the new instance derived from the predicted value y^* given by the baseline time-prediction model and a given relative error (RE). Therefore, for every given RE, an adjusted actual value can be derived from the predicted value through Eq. (4.6). Due to such one-to-one mapping between the relative error and the adjusted actual value, the PMF of the adjusted actual value can be derived from the PMF of the relative error. That is, an adjusted actual value has the same probability density as the relative error used to derive it.

Therefore, for every new prediction made by the baseline time-prediction model, we can derive the histogram/PMF of the adjusted actual value. For example, Fig. 4.7 shows the histogram/PMF of the adjusted actual value derived from the histogram/PMF of the relative error shown in Fig. 4.6 and the initial predicted value of 150 time units generated by the baseline time-prediction model.

Furthermore, once a new instance has been predicted by the baseline time-prediction model and evaluated by its actual value, it can be used to update the histogram/PMF of the relative error of the baseline time-prediction model. In this manner, the histogram/PMF of the relative error is generalized to all testing instances that the baseline time-prediction model has been applied to. In the following, we consider the case when the histogram/PMF derived from initial testing data is not dynamically updated by new instances.

4.3.2.3   Cumulative Density Function (CDF) of the Adjusted Actual Value

We can further obtain the cumulative distribution function (CDF) of the adjusted actual value for a predicted value. If the histogram/PMF of the relative error is not dynamically updated, then two instances that obtain the same initial predicted

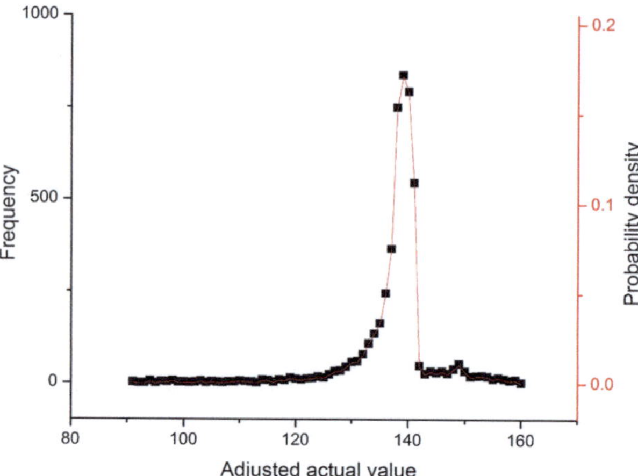

Fig. 4.7 The histogram and the probability mass function (PMF) of the adjusted actual value derived from the prediction given by the baseline time-prediction model and its histogram/PMF of the relative error shown in Fig. 4.6

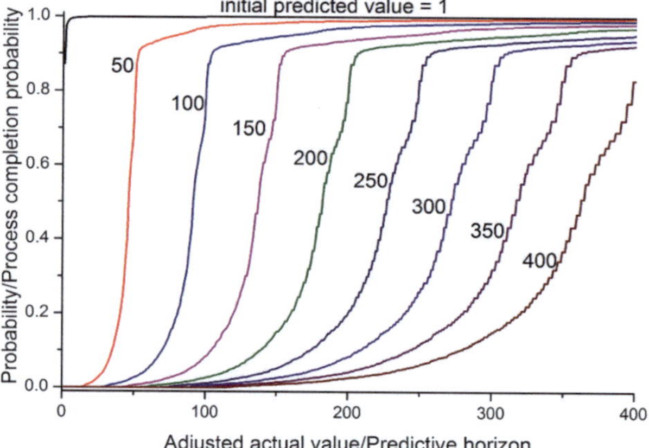

Fig. 4.8 The cumulative distribution functions (CDFs) of the adjusted actual value for different predicted values derived from the histogram/probability mass function (PMF) shown in Fig. 4.7

value from the baseline time-prediction model will have the same CDF of the adjusted actual value. Only for different predicted values, the CDF of the adjusted actual values are different. Figure 4.8 shows the CDFs of adjusted values for different initial predicted values. For any point on the CDF curve, the value on the vertical axis is the probability that the process-execution time is no greater than its corresponding adjusted actual value on the horizontal axis. Therefore, the value on the vertical axis is also the process completion probability by the predictive horizon value on the horizontal axis.

4.3.2.4   Expectation of the Actual Value

The proposed time-prediction method associates the validation data to the testing data by assuming that the validation data follows the same distribution as the testing data. Consequently, the relative error given by the baseline model follows the same

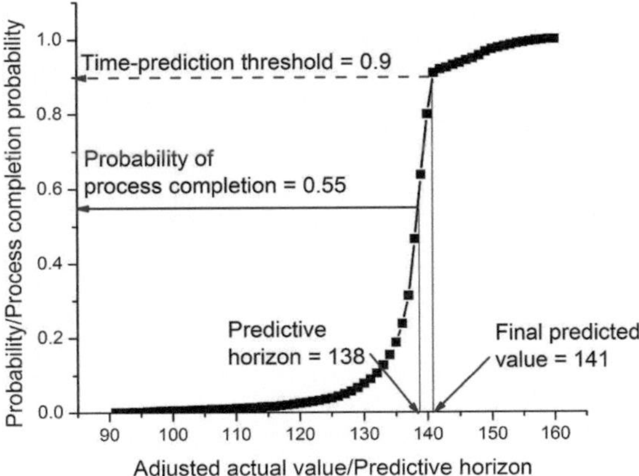

Fig. 4.9 The cumulative distribution function (CDF) of the adjusted actual value for one predicted value derived from the histogram/PMF shown in Fig. 4.7

distribution. Therefore, given a predicted value y^*, the expectation of the actual value can be obtained from Eq. (4.6) and the PMF of the relative error $f_{RE}(re)$ for the baseline time-prediction model as follows:

$$E[Y] = y^* \sum_{re} \frac{f_{RE}(re)}{1 + re} \tag{4.7}$$

Since $\sum_{re} f_{RE}(re) = 1$, $E[Y]$ is a weighted average of $y^*/(1 + re)$ that lies between $y^*/(1 + RE_{MAX})$ and $y^*/(1 + RE_{MIN})$.

4.3.2.5   Final Predicted Value

The final predicted process-execution time of the proposed time-prediction method is a candidate among the adjusted actual values. The CDF of the adjusted actual value is used for exploring the final predicted time. The proposed solution is to explore a time-prediction threshold and use its corresponding adjusted actual value on the CDF as the final predicted value. For instance, Fig. 4.9 shows the CDF of the adjusted actual value derived from the histogram/PMF shown in Fig. 4.7 and an initial predicted value equals 150 time units. If the time-prediction threshold is 0.9, then the final predicted value adjusted from the initial predicted value is 141 time units.

4.3.2.6   Time-Prediction Threshold

In order to make the most accurate process-execution time prediction, the time-prediction threshold should be optimized. Since the CDF of the adjusted actual value has a one-to-one mapping to the initial predicted value, the time-prediction threshold

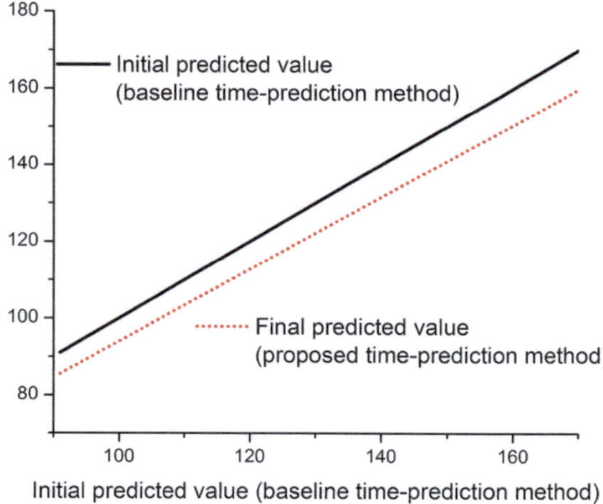

Fig. 4.10 The final predicted value derived from the initial predicted value under the condition that the optimal time-prediction threshold equals 0.94

should be separately optimized for each initial predicted value. Consequently, each initial predicted value corresponds to a specific final predicted value.

However, the time-prediction threshold optimization requires that each initial predicted value appears for a sufficient number of times. Even if a large number of testing instances can be evaluated, such assumption cannot be guaranteed. Therefore, we optimize a single time-prediction threshold to modulate all initial predicted values, such that the final predicted values make the proposed method to achieve both less relative mean error (RME) and root-mean-squared error (RMSE) than the initial predicted values.

Figure 4.10 demonstrates the one-to-one mapping of the initial predicted value to the final predicted value under the condition that the optimal time-prediction threshold equals 0.94.

4.3.3 Comparison Results and Discussions

The determination of an adequate sample size depends on the signal-to-noise ratio in the data, the nature of the problem to be solved, and a priori knowledge of whether the model specification is correct [35]. Figure 4.11 shows that in our demonstration test case the root-mean-squared error (RMSE) for the proposed method stabilize around 2,000 samples. The relative mean error (RME) has similar distribution pattern. Therefore, we kept at least 4,000 samples in each test case.

Table 4.3 shows the RME and RMSE tested through real EIS data by three methods: the baseline time-prediction method (only machine-learning), the expectation of the actual value (statistical analysis based on machine-learning result), and the proposed time-prediction method (machine-learning integrates statistical analysis). For instance, when evaluated using the demonstration test case, the RME and the

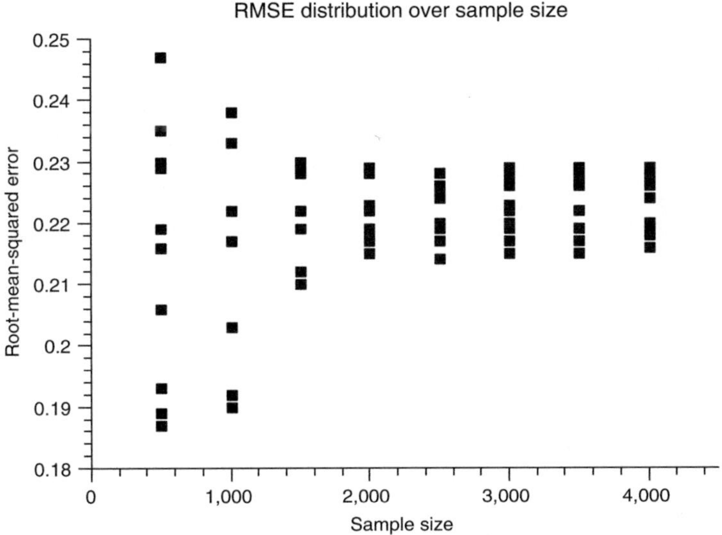

Fig. 4.11 The distribution of root-mean-squared error over sample size for the demonstration test case

Table 4.3 Relative mean error (RME) and root-mean-squared error (RMSE) of the baseline time-prediction method, the computed expectation of the actual value, and the proposed time-prediction method tested through real EIS data

Test case based on real enterprise data	Relative mean error			Root-mean-squared error		
	Baseline	Expectation of actual value	Proposed	Baseline	Expectation of actual value	Proposed
Demonstration	0.226	0.218	0.205	0.144	0.138	0.134
Other	0.324	0.309	0.293	0.234	0.231	0.223
	0.456	0.449	0.444	0.365	0.355	0.348
	0.123	0.113	0.098	0.102	0.102	0.096
	0.286	0.270	0.265	0.201	0.199	0.197
	0.334	0.309	0.302	0.267	0.257	0.248
	0.224	0.201	0.198	0.172	0.169	0.166

RMSE of the baseline time-prediction method are 0.226 and 0.144, respectively. The expectation of the actual value reduces the RME and the RMSE to 0.218 and 0.138, respectively. Furthermore, the proposed time-prediction method achieved smallest values of both RME and RMSE, 0.205 and 0.134, respectively. When evaluated using six other real test cases, the proposed time-prediction method has always achieved less RME and RMSE than the baseline time-prediction method. It has also achieved less RME and RMSE than the expectation of the actual value, which is the most common unbiased estimation of the actual value [36]. The time-prediction threshold is created in order to integrate machine learning method and statistical analysis. We can see that by optimizing the time-prediction threshold the proposed method can further improve the prediction accuracy.

The proposed time-prediction method generates the final prediction based on the performance of the baseline time-prediction method. For instance, if the baseline time-prediction method tends to overestimate (as shown in Fig. 4.6), then the proposed time-prediction method modulates the final predicted value to a smaller value through the time-prediction threshold and the CDF of the adjusted actual value (as shown in Fig. 4.10). In addition, the proposed time-prediction method also outperforms theoretical results since the learning-based time-prediction threshold learns deviations present in unseen validation data from theoretical values computed based on seen training and testing data.

4.4 Process Status Prediction

4.4.1 Baseline Status-Prediction Methods

There are two baseline status-prediction methods that we consider in this chapter. The first method shown in Fig. 4.12a compares the initial predicted process-execution time with the predictive horizon to check whether the process is still in progress. Such a method uses a regression model to realize a classifier. As long as time prediction is accurate enough, it can be further used to realize status prediction. If time prediction has a large error, then the error propagates to status prediction.

The second baseline status-prediction method shown in Fig. 4.12b builds individual status-prediction classification models for all possible predictive horizons. A status-prediction classification model is used to directly predict the process status. By replacing the numerical target value (i.e., the actual process-execution time) with the binary categorical target value (i.e., the actual process status at the given predictive horizon), a regression model can be easily converted into a classification model that is specifically built for the given prediction horizon. Clearly, the training

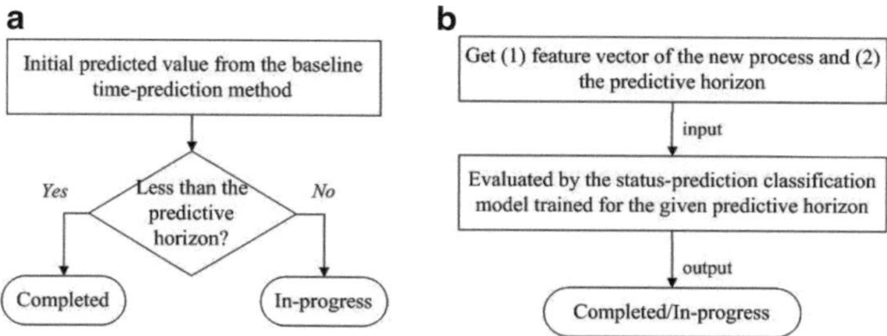

Fig. 4.12 Illustration of two baseline status-prediction methods base on (**a**) the baseline time-prediction regression model (**b**) the baseline status-prediction classification model

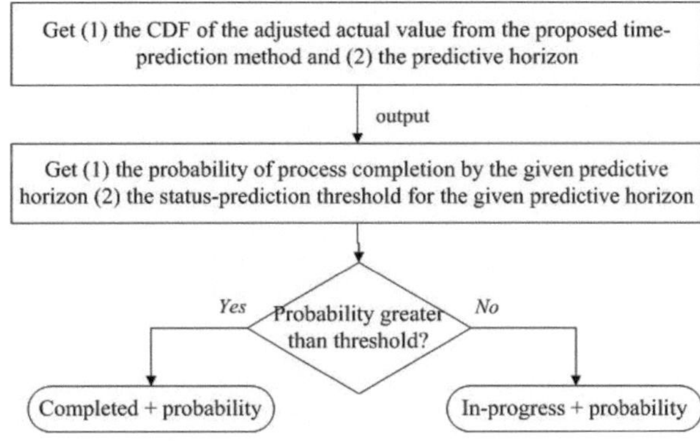

Fig. 4.13 Illustration of the proposed status-prediction method

process of the classification model requires the value of the predictive horizon to be known in advance to decide the target process status. Therefore, in order to predict process status for an arbitrary predictive horizon, one classification model has to be built for each predictive horizon. Such a feature makes the second baseline method computationally expensive due to the arbitrary nature of the predictive horizon.

4.4.2   Proposed Status-Prediction Method

Figure 4.13 shows the proposed status-prediction method. The detailed procedure is discussed as below.

4.4.2.1   Initial Prediction

Given an arbitrary predictive horizon, the probability of process completion by the predictive horizon can be directly obtained from the CDF of the adjusted actual value. It equals the probability that the process execution time is no greater than the predictive horizon. For instance, as shown in Fig. 4.9, if the predictive horizon is 138 time units, then the probability of process completion by 138 time units once it has been started is 0.55. The CDF also shows that it is highly unlikely that the process can be completed within 90 time units and will most likely be completed after 160 time units once it has been started.

4.4.2.2   Conditional In-Progress Prediction

Furthermore, according to probability theory [37], if a process has already been started and is still in progress, the probability of process completion by a predictive horizon is the conditional probability.

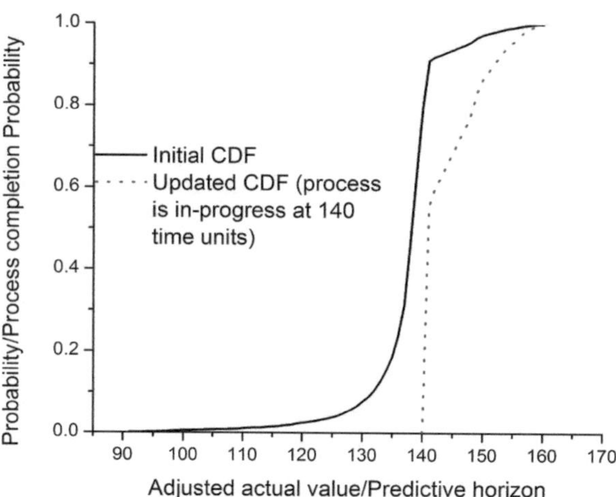

Fig. 4.14 The updated cumulative distribution function (CDF) of the adjusted actual value derived from the histogram/probability mass function (PMF) shown in Fig. 4.9 and the evidence that the process is in progress at time 140

$$P(\text{completed by } t_2 | \text{in-progress at } t_1)$$

$$= \frac{P(\text{completed between } t_1 \text{ to } t_2)}{P(\text{completed after } t_1)} \tag{4.8}$$

Equation (4.8) shows that the probability of process completion by time t_2 given the evidence that the process is still in-progress at time t_1 equals the probability of process completion between time t_1 and t_2 divided by the probability of process completion after time t_1. The CDF of the adjusted actual value should also be updated by the given evidence that the process is still in-progress at time t_1. For example, Fig. 4.14 shows the updated CDF of the adjusted actual value by Eq. (4.8) from Fig. 4.9, given that the process is still in-progress by 140 time units.

4.4.2.3   Status-Prediction Threshold

As demonstrated through Figs. 4.9 and 4.14, given a predictive horizon, the probability of process completion can be found through the CDF of the adjusted actual value. We define a status-prediction threshold in order to determine the class label as shown in Fig. 4.13. By comparing the probability of completion to the status-prediction threshold, the final class label is "completed" if the probability is greater than the threshold, and "in-progress" if it is lower than the threshold.

The probability of process completion varies with the predictive horizon. Intuitively, the larger the predictive horizon, the higher the probability that the predicted status is "completed". Therefore, the optimal status-prediction threshold that results in the minimal *error rate* status-prediction classifier should vary with the predictive horizon.

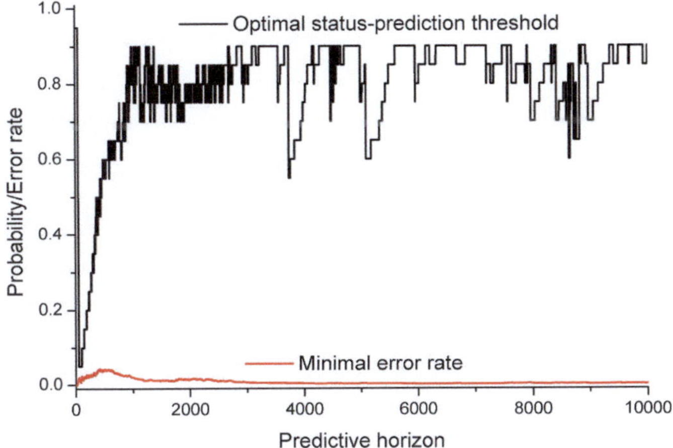

Fig. 4.15 Variations of the optimal status-prediction threshold and the *error rate* with respect to the predictive horizon

Figure 4.15 is obtained from optimizing the status-prediction thresholds for the demonstration test case. It shows that both the status-prediction threshold and the minimal *error rate* varies with the predictive horizon. Therefore, for a certain predictive horizon, the status-prediction model should select its corresponding status-prediction threshold for making the final classification.

Different from optimizing the proposed time-prediction method (discussed in Sect. 4.3.2.6, in which the time-prediction threshold can be optimized with respect to the initial predicted value generated by the baseline time-prediction model), the optimal status-prediction threshold of the proposed status-prediction method is explored with respect to the predictive horizon given by the model user.

4.4.3   Comparison Results and Discussions

We evaluate the performance of the proposed status-prediction method compared to the baseline methods. In the proposed method, we compare the process-completion probability (obtained from the CDF of the adjusted actual value, derived in Sects. 4.4.2.1 and 4.4.2.2) with the optimal status-prediction threshold of the predictive horizon (derived in Sect. 4.4.2.3). In the first baseline method, the process status is determined by comparing the initial predicted value with the predictive horizon.

Figure 4.16a shows that the performance of the proposed method is always superior than the first baseline method regardless of the predictive horizon. Figure 4.16b further zooms into the predictive horizon ranged from 230 to 590 time units. Both of them show that the *error rate* varies with respect to the predictive horizon.

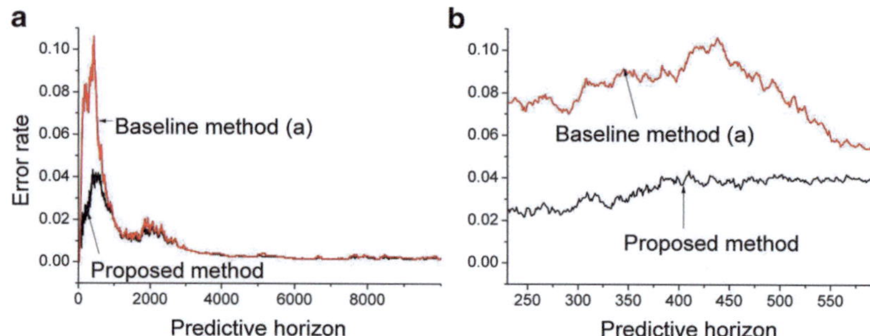

Fig. 4.16 The error rate of the proposed status-prediction method and the first baseline status-prediction method with respect to the predictive horizon ((**b**) is a zoom-in version of (**a**))

Table 4.4 *Error percentage results of the two baseline status-prediction methods (Fig. 4.12) and the proposed status-prediction method at some sampled predictive horizons*

| Predictive horizon (time units) | *Error percentage (%)* | | Proposed method |
| | Baseline methods | | |
	(a)	(b)	
100	5.859	5.486	1.818
150	7.525	7.563	2.071
200	7.879	7.623	2.121
250	7.475	7.195	2.576
300	7.626	7.236	2.879
350	8.939	7.852	2.929
410	9.899	8.536	4.343
450	9.343	8.125	3.939
500	8.081	8.227	4.091
550	5.909	5.365	3.939

As discussed in Sect. 4.4.1, the second baseline status-prediction method requires building a status-prediction classifier for each predictive horizon. Therefore, we tested the second baseline method with some sampled predictive horizons. Comparison results are shown in Table 4.4. For instance, it shows that when the predictive horizon equals to 410 time units (i.e., the predictive horizon for which the proposed status-prediction method gives the highest *error rate* as shown in Fig. 4.16), the *error rate* of the proposed method, the first baseline status-prediction method (a), and the second baseline status-prediction method (b) is 4.343 %, 9.899 %, and 8.536 % respectively. It can be seen that the proposed status-prediction method demonstrates better performance than both the baseline status-prediction methods.

Furthermore, we have tested other real test cases, our proposed method always achieve higher accuracies in both process-execution time (Table 4.3) and process status predictions. The improvement of the proposed status-prediction method over the baseline status-prediction methods is mainly due to the improvement of the proposed time-prediction method over the baseline time-prediction method (discussed in Sect. 4.3.3). In addition, the status-prediction threshold takes into consideration the impact of the arbitrary nature of the predictive horizon. Status-prediction threshold serves as a new tunable parameter that can further enhance prediction accuracy.

4.5 Conclusion and Future Work

In this chapter, we have developed a framework for process-execution time and status predictions, and compared results to baseline (state-of-the-art) approaches using real EIS data.

We have shown that, although machine-learning algorithms have demonstrated their successes in time prediction, little work has been done to further explore its capability and flexibility. The proposed models extend baseline models and have achieved better performance as well as new functions and more flexibility.

The advantages of the proposed models and the reasons for achieving superior performance than baseline approaches can be summarized as follows:

- By integrating statistical analysis and machine-learning algorithms, the proposed time-prediction model takes into consideration the performance deviation of the baseline time-prediction model with respect to different instances. It reasonably adjusts the initial predicted value to get the final predicted value such that less relative mean error and less root-mean-squared error can be achieved. The proposed model also absorbs the effect of outliers as long as its histogram of the relative error is derived from a sufficient number of testing instances.
- Along with a binary status prediction, the proposed process status prediction model also provides a probability estimation of the predicted status. It further enables conditional predictions if the intermediate status becomes known as evidence.
- By replacing multiple classification models by a single regression model, our status-prediction model greatly reduces the cost of building extra models. It also considers the effects caused by the arbitrary predictive horizon.

Finally, in our future work, the performance of the proposed system will be tested through a discrete-event simulation platform as an embedded artificial intelligent functional block in the whole EIS.

References

1. A.-W. Scheer, M. Nüttgens, *ARIS Architecture and Reference Models for Business Process Management* (Springer, 2000)
2. C.L. Dunn, J.O. Cherrington, A.S. Hollander, E.L. Denna, *Enterprise Information Systems: A Pattern-Based Approach*, vol. 3 (McGraw-Hill/Irwin Boston, 2005)

3. Q. Duan, J. Zeng, K. Chakrabarty, G. Dispoto, Real-time production scheduler for digital-print-service providers based on a dynamic incremental evolutionary algorithm. Accepted for publication in IEEE Trans. Autom. Sci. Eng. Available on IEEE Xplore as Early Access article, vol. PP, no. 99, pp. 1–15 (2014)

4. K. Chakrabarty, G. Dispoto, R. Bellamy, J. Zeng, The role of EDA in digital print automation and infrastructure optimization, in *Proceedings of the International Conference on Computer-Aided Design (ICCAD)* (2011), pp. 158–161

5. D.H. Kim, S. Mukhopadhyay, S.K. Lim, TSV-aware interconnect distribution models for prediction of delay and power consumption of 3-d stacked ICs. IEEE Trans. Comput.-Aided Des. Integr. Circuits Syst. **33**(9), 1384–1395 (2014)

6. E.-J. Chang, H.-K. Hsin, S.-Y. Lin, A.-Y. Wu, Path-congestion-aware adaptive routing with a contention prediction scheme for network-on-chip systems. IEEE Trans. Comput.-Aided Des. Integr. Circuits Syst. **33**(1), 113–126 (2014)

7. P. van Stralen, A. Pimentel, Fitness prediction techniques for scenario-based design space exploration. IEEE Trans. Comput.-Aided Des. Integr. Circuits Syst. **32**(8), 1240–1253 (2013)

8. R. Schneider, D. Goswami, S. Chakraborty, U. Bordoloi, P. Eles, Z. Peng, Quantifying notions of extensibility in flexray schedule synthesis. ACM Trans. Des. Autom. Electron. Syst. **19**(4), 32:1–32:37 (2014)

9. A. Tenhiälö, M. Ketokivi, Order management in the customization-responsiveness squeeze. Decis. Sci. **43**(1), 173–206 (2012)

10. A. Scheer, F. Habermann, Enterprise resource planning: making ERP a success. Commun. ACM **43**, 57–61 (2000)

11. W. Lee, Y. Wang, D. Shin, N. Chang, M. Pedram, Optimizing the power delivery network in a smartphone platform. IEEE Trans. Comput.-Aided Des. Integr. Circuits Syst. **33**(1), 36–49 (2014)

12. S. Biswas, H. Wang, R.D.S. Blanton, Reducing test cost of integrated, heterogeneous systems using pass-fail test data analysis. ACM Trans. Des. Autom. Electron. Syst. **19**(2), 20:1–20:23 (2014)

13. D.-C. Juan, S. Garg, D. Marculescu, Statistical peak temperature prediction and thermal yield improvement for 3d chip multiprocessors. ACM Trans. Des. Autom. Electron. Syst. **19**(4), 39:1–39:23 (2014)

14. H. Kipphan, *Handbook of Print Media: Technologies and Production Methods* (Springer, New York, 2001), pp. 40–422

15. M.A. Hall, Correlation-based feature selection for machine learning. Ph.D. dissertation, The University of Waikato (1999)

16. J.D. Musa, K. Okumoto, A logarithmic poisson execution time model for software reliability measurement, in *Proceedings of the 7th International Conference on Software Engineering*, ser. ICSE '84 (IEEE Press, Piscataway, 1984), pp. 230–238

17. P. Bogdan, M. Kas, R. Marculescu, O. Mutlu, Quale: a quantum-leap inspired model for non-stationary analysis of noc traffic in chip multi-processors, in *Proceedings of the 2010 Fourth ACM/IEEE International Symposium on Networks-on-Chip*, ser. NOCS '10 (IEEE Computer Society, Washington, DC, 2010), pp. 241–248

18. Z. Qian, P. Bogdan, G. Wei, C.-Y. Tsui, R. Marculescu, A traffic-aware adaptive routing algorithm on a highly reconfigurable network-on-chip architecture, in *Proceedings of the Eighth IEEE/ACM/IFIP International Conference on Hardware/Software Codesign and System Synthesis*, ser. CODES+ISSS '12 (ACM, New York, 2012), pp. 161–170

19. M. Xue, K.K. Droegemeier, V. Wong, The advanced regional prediction system (ARPS) – a multi-scale nonhydrostatic atmospheric simulation and prediction model. Part i: model dynamics and verification. Meteorol. Atmos. Phys. **75**(3–4), 161–193 (2000)

20. R.M. Fujimoto, Parallel and distributed simulation, in *Simulation Conference Proceedings, 1999 Winter*, vol. 1 (IEEE, 1999), pp. 122–131

21. C. Gopinath, J.E. Sawyer, Exploring the learning from an enterprise simulation. J. Manage. Dev. **18**(5), 477–489 (1999)

22. F. Darema, Dynamic data driven applications systems: a new paradigm for application simulations and measurements, in *Computational Science-ICCS 2004* (Springer, 2004), pp. 662–669
23. Z. Qian, D.-C. Juan, P. Bogdan, C.-Y. Tsui, D. Marculescu, R. Marculescu, Svr-noc: a performance analysis tool for network-on-chips using learning-based support vector regression model, in *Proceedings of the Conference on Design, Automation and Test in Europe*, ser. DATE '13 (EDA Consortium, San Jose, 2013), pp. 354–357
24. D. Tetzlaff, S. Glesner, Intelligent prediction of execution times, in *2013 Second International Conference on Informatics and Applications (ICIA)*, Sept 2013, pp. 234–239.
25. S. Ravizza, J. Chen, J.A. Atkin, P. Stewart, E.K. Burke, Aircraft taxi time prediction: comparisons and insights. Appl. Soft Comput. **14**(Part C, 0), pp. 397–406 (2014)
26. R.J. Klement, M. Allgäuer, S. Appold, K. Dieckmann, I. Ernst, U. Ganswindt, R. Holy, U. Nestle, M. Nevinny-Stickel, S. Semrau, F. Sterzing, A. Wittig, N. Andratschke, M. Guckenberger, Support vector machine-based prediction of local tumor control after stereotactic body radiation therapy for early-stage non-small cell lung cancer. Int. J. Radiat. Oncol*. Biol*. Phys. (2014)
27. C.-H. Wu, J.-M. Ho, D. Lee, Travel-time prediction with support vector regression. IEEE Trans. Intell. Transp. Syst. **5**(4), 276–281 (2004)
28. J. Zeng, S. Jackson, I. Lin, M. Gustafson, E. Gustafson, R. Mitchell, Operations simulation of on-demand digital print, in *IEEE 13th International Conference on Computer Science and Information Technology* (2011)
29. A. Simitsis, P. Vassiliadis, T. Sellis, Optimizing etl processes in data warehouses, in *Proceedings of the 21st International Conference on Data Engineering* (2005), pp. 564–575
30. F. Ye, Z. Zhang, K. Chakrabarty, X. Gu, Board-level functional fault diagnosis using multi-kernel support vector machines and incremental learning. IEEE Trans. Comput.-Aided Des. Integr. Circuits Syst. **33**(2), 279–290 (2014)
31. F. Ye, Z. Zhang, K. Chakrabarty, X. Gu, Board-level functional fault diagnosis using artificial neural networks, support-vector machines, and weighted-majority voting. IEEE Trans. Comput.-Aided Des. Integr. Circuits Syst. **32**(5), 723–736 (2013)
32. Q. Duan, Real-time and data-driven operation optimization and knowledge discovery for an enterprise information system. Ph.D. dissertation, Duke University, Department of Electrical and Computer Engineering, Durham, 2014
33. J.C. Platt, Sequential minimal optimization: a fast algorithm for training support vector machines. Technical report, Microsoft Research (1998)
34. M. Hall, E. Frank, G. Holmes, B. Pfahringer, P. Reutemann, I.H. Witten, The weka data mining software: an update. SIGKDD Explor. **11**(1) (2009). Available: http://www.cs.waikato.ac.nz/ml/weka/
35. G.J. Huffman, Estimates of root-mean-square random error for finite samples of estimated precipitation. J. Appl. Meteorol. **36**(9), 1191–1201 (1997)
36. P.R. Halmos et al., The theory of unbiased estimation. Ann. Math. Stat. **17**(1), 34–43 (1946)
37. K. Trivedi, *Probability and Statistics with Reliability, Queuing and Computer Science Application*, 2nd edn. (Wiley, New York, 2001)

Chapter 5
Optimization of Order-Admission Policies

A mass-customization enterprise offers personalized manufacturing services. Once an order is submitted to the enterprise by a client, the EIS needs to make a real-time decision on whether to accept or refuse this order. Based on the enterprise current capacity, and the order's properties and requirements, an order is refused if its acceptance is not profitable for the enterprise. The order is accepted with the most appropriate due date in order to maximize the profit that can result from this order. We have developed an intelligent order-admission framework that provides admission decisions in real-time for new orders using machine-learning and decision-integration techniques. The framework consists of three classifiers: Support Vector Machine (SVM), Decision Tree (DT), and Bayesian Probabilistic Model (BPM). The classifiers are trained by history orders and used to predict completion status for new orders. A decision integration technique is implemented to combine the results of the classifiers and predict due dates. Experimental results derived using real factory data from a leading print-service provider and Weka open source software show that the order completion-status prediction accuracy is significantly improved by the decision-integration strategy. The proposed multi-classifier model also outperforms a stand-alone regression model.

This chapter addresses order-admission policy optimization in an EIS using digital-print industry as an example. The rest of this chapter is organized as follows. Section 5.1 discusses prior art and motivation. An order-fulfillment prediction model which includes multiple classification models has been built to estimate order life cycle. The main components and the structure of the model are discussed in Sect. 5.2. Section 5.3 describes decision integration strategies for increasing the accuracy of model. In Sect. 5.4, we present simulation results and discuss the performance of the model. Section 5.5 concludes the chapter.

© Springer International Publishing Switzerland 2015

Q. Duan et al., *Data-Driven Optimization and Knowledge Discovery for an Enterprise Information System*, DOI 10.1007/978-3-319-18738-9_5

5.1 Background and Motivation

A real-time embedded system within an operational system can be viewed as an intelligent component for a dedicated function [1]. For system-wide design and optimization, time-consumption prediction is crucial in almost any operational environment. For example, software optimization requires time-consumption estimation for each software stage [2], yield optimization for embedded systems-on-chip requires time-consumption information for each resource [3], and communication system design and optimization requires message-response time [4]. Moreover, for real-time embedded systems, accurate estimates of completion times are preferred compared to simple worst-case execution times [5, 6]. Traditionally, pre-designed rule-based specifications or even human judgment were sufficient for this purpose. However, increased system complexity and heterogeneity motivates the need for advanced algorithms for real-time performance prediction in embedded systems [1]. A typical example of such a real-time embedded system is a commercial digital print enterprise.

Commercial print is a highly competitive and service quality-oriented business with annual retail sales of over US$700B [7, 8]. Establishing a superior service engagement and fulfillment between clients and the print service providers (PSPs) is key to creating value and ensuring profit for a commercial print business [9]. The end-to-end fulfillment process of digital-print service starts with clients' engagement to obtain print services [10]. The clients, who are also the content suppliers, request print services and supply content for print. Clients range from traditional publishing agencies for newspapers, books, and periodicals, to web-based digital media companies, e.g., *www.snapfish.com*. The PSP offers its manufacturing capability to the content suppliers as a form of utility service in exchange for payment. Today, on-demand digital-print services are leading the digitization of commercial print and commanding an impressive 8 % annual growth rate [7]. Compared with traditional print services, which produce homogeneous print products (thousands of copies of the same content, also known as "long run-length"), on-demand digital-print deals with high-frequency and high-volatility service demands [11].

In the commercial digital-print industry, customers can request print products using multiple methods. They may physically visit the PSP or electronically submit print content to the PSP through the internet. In order to describe a potential order, a document called "Statement of Work" (SoW) is created which provides three crucial pieces of information. The first is order intent, which specifies the customer's print intent, for example, the types of products, the number of copies, as well as particular instructions such as laminating and quality requirements. The second is the expected due date which can be proposed from two methods. It can be requested by the client or given by the PSP. The third is the price of this order, which similarly, may come from either the client's offer or the PSP's proposal.

After a negotiation process, if the order is accepted, the price and the due date are finalized and a contract for this order is established. A production plan that specifies all the production steps to fulfill this order, is also devised and disclosed to the client. The production steps include sequences of production operations on the factory floor, including pre-press production, cover making, book block printing, laminating, cutting, collating, binding, packaging and shipping, etc. [12]. Besides offering high quality and good price, the PSP must deliver an order by its deadline since late delivery leads to extra shipping cost and reduces customer satisfaction or loyalty.

The most popular practice in today's commercial print industry for order acquisition is based on human involvement [7]. This kind of contract establishment process between a PSP and the clients mainly depends on human judgment. When a potential service engagement arrives at a PSP, the production managers and the customer service representatives cooperate and then present to the client a tentative price and completion date. Such an approach is clearly incompatible with today's mass customized production due to increasing diversity and complexity [11].

5.1.1   Related Prior Solutions

A more advanced technique implemented by a few of the industry-leading PSPs (e.g., Reischling Press, Inc. (RPI)), is a template-based automated order acquisition process [12]. First, a set of pre-determined static templates are developed through business cooperation with a strategic print buyer over several months. Each template represents a type of product and includes the associated rules to calculate the lead time of a product that can be mapped into this template. Once the templates are well established, they can be deployed so that the order acquisition process can be automated by using them. A new set of templates is required to be developed once the situation in the factory changes. Figure 5.1 demonstrates the template-based lead-time estimation process for an order involving two products. Other examples of similar approaches include a case-based reasoning system for predicting job completions [13], and a fuzzy rule-based prediction of monthly precipitation [14].

None of the above solutions can account for the up-to-the-moment factory production situation. These techniques are static, non-adaptive and susceptible to faults. Smith [15] has discussed the difficulty associated with decision-making in practical production management domains and identified the general opportunities that data-driven artificial intelligence systems can provide. For a real-life problem, [16] demonstrated that a data-driven machine-learning solution was needed when simple matching based on observations became ineffective for detecting target signals. In [17], memory based reasoning (MBR) and neural network (NN) learning are combined for yield improvement and an integrated framework is proposed for a yield management system based on hybrid machine-learning techniques. In [18], a set of heterogeneous independent learning components were coordinated by a central meta-reasoning executive. It has been demonstrated that machine-learning techniques assist us to manage complexity, diversity, and uncertainties in manufacturing process management [19].

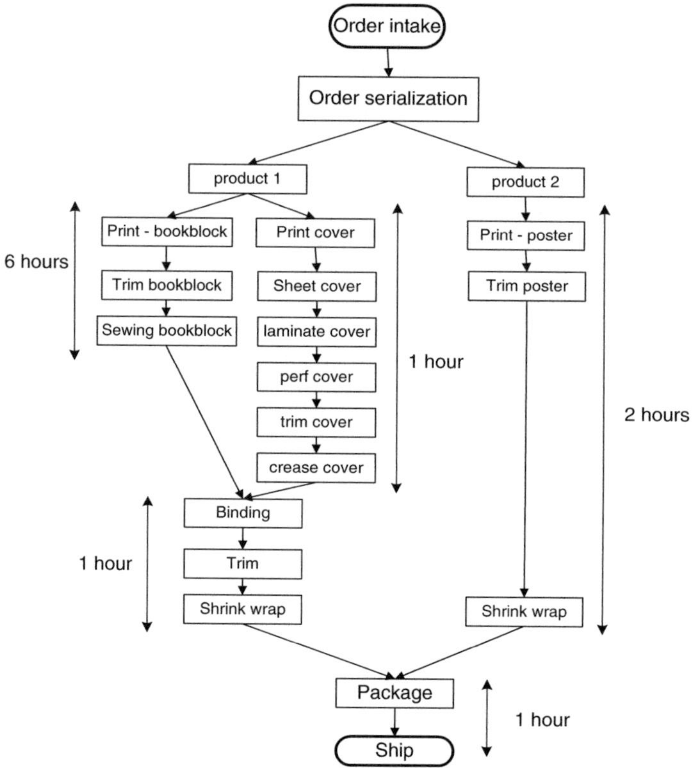

Fig. 5.1 An example of a production process for a two-product order

5.1.2 Costs for Service-Level Violation

Print production suffers from high process variation due to nontrivial occurrence of exception events that are beyond the planned production process. Typical examples are product reworking induced by machine malfunctions, digital content repair due to inconsistent files, service maintenance and replenishment of the consumables (paper rolls, inks, etc.). All these exception events negatively impact the realization of the due-date commitment. In general, if an order cannot be completed on time, the PSPs may incur extra costs, for instance:

- Cost to upgrade shipping methods. This common remedial action for PSPs is used for compensating for the extra time consumed by manufacturing. Such unexpected shipping cost can be significant; it can reach US$0.5M through a busy season [11].

- Cost of goodwill. This cost is client-dependent. It is particularly important when PSPs conduct business with enterprise print buyers. A significant delivery delay beyond the promised due date given by the PSP can jeopardize the enterprise's own business objective. Negative service experience consequently hurts a PSP's re-occurring contracts and business development effort.
- Possible late penalty stated in the service contract.

On the other hand, PSPs can choose to reduce the price from what has been advertised if the customer agrees on a postponed due date. This type of cost is the price difference between the reduced price caused by postponing the due date and the original price. In this scenario, a postponed due date increases the likelihood of timely delivery, therefore, PSPs can accept the price penalty in order to avoid the potentially higher amount of late cost.

The cost to upgrade shipping methods can be quantified using the cost table provided by the shipping companies; the cost depends on the size (weight, volume) of the print product and the reduction of the shipping days. The cost of goodwill can be quantified by examining the Customer Relation Management (CRM) database and drawing correlations between the previous late service events and lost business opportunities. The cost stated in the service contract can be quantified by analyzing the service contract. The cost can also be quantified by examining the history of postponed service events.

Most factory-optimization problems can be viewed as being composed of two elements: a due-date setting element, and a scheduling element. Indeed, the interplay between scheduling orders to meet due dates and the proper setting of due dates makes difficult or even impossible to do co-optimization in a heterogeneous and stochastic factory environment [20]. The approach presented in this chapter is a due-date-based model in which the due date is first assigned, and then the schedule is determined based on order due dates. We cannot fit any standard models considered in [20] (e.g., single-machine common-due-date models, single-machine static-distinct-due-date models, single-machine dynamic models, parallel-machine models, and jobshop and flowshop models) to our real-world factory environment, which is characterized as being transient, heterogeneous, and stochastic. Therefore, even the simplified scheduling model is at least as hard as known NP-complete problems. This problem is addressed in more detail in [21], which considers scheduling and resource allocation for a digital print factory.

The scheduling strategy is treated as a black box in this work. We do not consider scheduling in this chapter. In fact, the scheduling problem in the context of digital print enterprises was considered in [21]. The proposed prediction system and classical real-time OS scheduling techniques can therefore be used together to generate even better prediction results. Note also that classical OS scheduling strategies do not consider prediction or machine learning to learn from data. The aim of this study is to predict whether an order can be completed within a proposed due date under the scheduling strategy. We introduce a new order-admission framework and show the efficiency of data-driven approaches adopted by the framework. The key contributions of this chapter are listed below:

- Relevant order- and factory-information attributes have been identified and used in a classification framework to make predictions for new orders.
- Decision integration techniques have been used to improve the prediction accuracy.
- Order due dates are explored through order completion-status classification results.
- We show that the proposed order-admission framework reduces the number of orders that miss their due dates.

The rest of this chapter is organized as follows. We describe the main components of the classification framework and the relationship between them in Sect. 5.2. Section 5.3 describes decision integration strategies for increasing the accuracy of classification. In Sect. 5.4, we present experimental results derived using the Weka, open source software [22], and discuss the performance of the proposed framework. Section 5.5 concludes the chapter.

5.2   Due-Date Validation Engine

As shown in Fig. 5.2, whenever a new order arrives at the factory, relevant information about this order along with current factory information is sent to the due-date validation engine. Based on the initial due date required by the customer, the due-date validation engine makes a prediction whether it can be delivered on time. If it can be completed as expected, then the contract can be established between the PSP and the customer. If it cannot be completed as expected based on current factory status, then the due-date validation engine proposes a new due date. The customer can decide whether the initial due date can be postponed. If not,

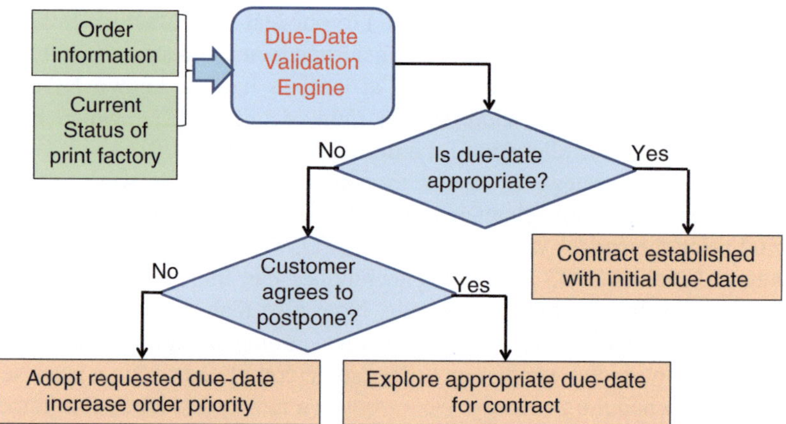

Fig. 5.2 Illustration of the proposed order-admission framework

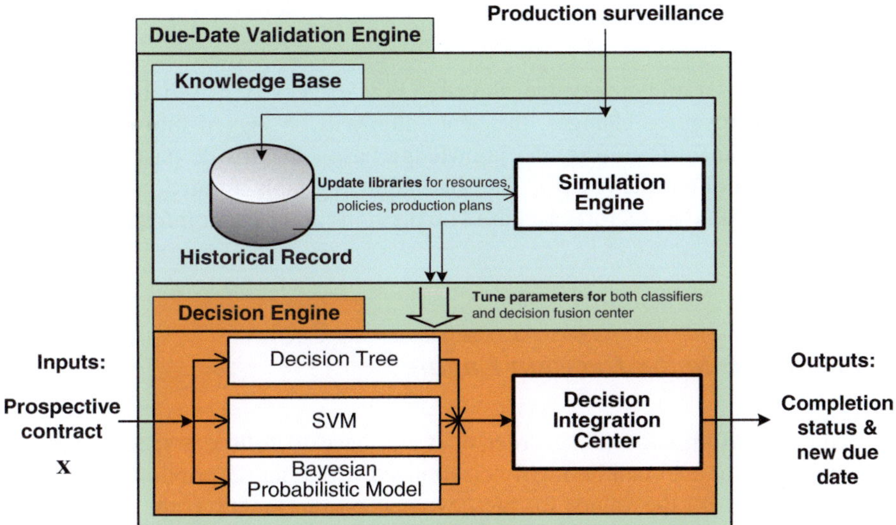

Fig. 5.3 System architecture and activity diagram of the due-date validation engine

the customer can pay more to increase the priority of this order. High-priority orders take less time to be completed since they are also scheduled in advance compared to regular orders.

The due-date validation engine is designed to automate the order admission process. The functionality of the engine is to output an instantaneous response to the contract on whether it can be delivered on time, and if not, the most appropriate due date. Figure 5.3 shows the schematic of the due-date validation engine. The information of every incoming prospective contract is fed into the engine before a contract can be established.

The main difficulty in achieving high predictability is caused by two factors. One factor is the uncertainty that is inherent in the arrival of new orders. The other factor is the inevitably insufficient amount of input information. The available information for prediction at the moment when an order arrives only includes the history and the up-to-the-moment factory status information. However, whether an order can be completed on time is also affected by the factory status during the manufacturing process, which is unknown at the time when the prediction must be made. In order to address the first obstacle, instead of relying on just one type of machine learning algorithm, we search for different algorithms to perform the prediction such that a broader coverage can enhance the overall prediction accuracy. It has been shown that integrating the advantages of multiple machine-learning algorithms leads to higher efficiency in practical scenarios [23, 24]. The second problem can be overcome by forecasting the factory performance, which is beyond the scope of this chapter and left for future research. The key components and functions of the proposed due-date validation engine are explained in detail in the following subsections.

5.2.1   Knowledge Base

The production surveillance system shown in Fig. 5.3 is event driven, i.e., whenever a feature of a resource changes, the new value will be logged into databases in a real-time manner. Therefore, the knowledge base can provide past and current information about the factory [8]. The simulation engine is a functional block that can provide factory status forecast based on previous and current information about the factory.

5.2.2   Inputs to the Decision Engine

The contract information of an order does not have to follow a specific format. Contents of a contract can be highly customized. Contract descriptions are parsed and extracted to fill the values for pre-designed system inputs. If the value for an input cannot be mined from a contract description, it corresponds to a missing value. Machine-learning algorithms are able to handle missing values, e.g., a missing value can be replaced by a median or a default value.

An attribute vector $\mathbf{x}$, which consists of 14 carefully selected key attributes from both the contract and the factory is constructed for every prospective contract. The same vector $\mathbf{x}$ is sent to all classifiers each time. Critical contract attributes are identified from the "Statement of Work". The key factory status information is supplied by the knowledge base once the contract arrives at the factory. An example input vector $\mathbf{x}$ is shown as $\mathbf{x} = (A_1, A_2, \ldots, A_{14})$, and each attribute is explained in Table 5.1. A value stream is a sequence of processes required to produce a specific type of product. For example, value stream "Poster" is the rightmost template shown in Fig. 5.1.

Leaving out relevant attributes or keeping irrelevant attributes may be detrimental for knowledge discovery, causing confusion for the machine-learning algorithm employed and resulting in discovered patterns of poor quality. In addition, the added irrelevant or redundant attributes can slow down the model building process [25]. There were more than 14 attributes available in our problem, but many of these were found to be irrelevant to the classification model or redundant. The final 14-attribute subset was obtained by using the heuristic variable subset selection method discussed in [26] to remove attributes that did not improve accuracy or even led to a deterioration of the model-prediction accuracy. It can be seen that the final selected attributes include both order-property attributes (A_1-A_{11}) and factory-status attributes (A_{12}-A_{14}) that have an effect on order lead time.

The factory complexity is correlated to the prediction models. We considered as many inputs as possible as discussed above to address job complexity. In the feature-selection phrase, we retained only 14 predictive inputs based on our assessment of their importance. Therefore, the more complex the system, the more inputs there are to the decision engine.

Table 5.1 Attributes of the input vector

Index	Attribute	Explanation
A_1	*OrderLeadTime*	The proposed due date minus the accepted time
A_2	*ProductQuantity*	The number of products in this order
A_3	*BookBlockQuantity*	The number of book blocks in this order
A_4	*BookCoverQuantity*	The number of book covers in this order
A_5	*PerfectboundCover*	Indicates whether value stream "Perfectbound Cover" is required
A_6	*PerfectboundBook*	Indicates whether value stream "Perfectbound Book" is required
A_7	*CustomCover*	Indicates whether value stream "Custom Cover" is required
A_8	*CustomBook*	Indicates whether value stream "Custom Book" is required
A_9	*Poster*	Indicates whether value stream "Poster" is required
A_{10}	*StandardBook*	Indicates whether value stream "Standard Book" is required
A_{11}	*CustomWrap*	Indicates whether value stream "Custom Wrap" is required
A_{12}	*WorkInProgress*	The number of products that are currently in the factory
A_{13}	*ResourceAQueueLength*	The number of tasks waiting in the queue of book-block presses
A_{14}	*ResourceBQueueLength*	The number of tasks waiting in the queue of book-cover presses

5.2.3   *Outputs of the Decision Engine*

Classifiers in the decision engine shown in Fig. 5.3 have been chosen because they are considered to be among the top-10 most influential machine-learning algorithms [27]. Each produces a prediction about the prospective contract's completion status and a probability estimation *Prob*. There are two possible completion status that can be viewed as two class labels. The "On-time order" label means that the factory is confident of delivering this order on time, and "Late order" implies that this order will potentially be late if it is due by the proposed due date. The probability estimate *Prob* is the confidence in making the prediction. Therefore, the output space **Y** for classifiers is a two-element set of class labels, **Y** = {"On-time order", "Late order"}. The probability estimate $Prob \in [50\,\%, 100\,\%]$. For example, if the prediction is "On-time order", then $Prob = Prob(N)$ and $Prob(N) + Prob(P) = 1$, where $Prob(N)$ and $Prob(P)$ are the probability estimations that an order is "On-time" or "Late" order.

5.2.4   *Classifier Evaluation Metrics*

Predictive accuracy is the main evaluation criterion for the predictive performance of classifiers [28]. There are four important metrics used for evaluating a classifier's performance and only two of them are independent of each other [29]. They are true positive rate (*TP*), true negative rate (*TN*), false negative rate (*FN*), and false positive rate (*FP*). They represent the rate of correctly classifying a late order as "Late order", correctly classifying an on-time order as "On-time order", falsely classifying a late order as "On-time order", and falsely classifying an on-time order as "Late order", respectively. They are related as follows:

$$TP + FN = 1 \tag{5.1}$$

$$TN + FP = 1 \tag{5.2}$$

Note that, classification is done at the moment when an order arrives; however, the actual class label is known only after the order is completed.

5.2.5   Support Vector Machines

A nonlinear two-class Support Vector Machine (SVM) is used as one classifier. An SVM is a supervised machine-learning algorithm [30]. It can be trained for solving binary classification problems. Although similar to neural networks [31], the quality of generalization and the ease of training exceed that of traditional methods. An SVM performs well on datasets that have many attributes. Even if very few training instances are available, as long as they are representative in the problem space, an SVM can still perform well. In addition, there is no upper limit to the number of attributes; hardware capabilities are the only constraints. SVMs have been successfully applied to model complex and real-world problems such as text and image classification, hand-writing recognition, as well as bioinformatics and biosequence analysis [32, 33]. A brief introduction to an SVM, illustrated by our problem, is presented below.

Each training instance corresponds to a history order. The training instance of an order i consists of an attribute vector $\mathbf{x}_i$ and a "target value" y_i (i.e., the class label, either "On-time order" or "Late order"). The goal of an SVM is to generate a model from training instances. The model can then make predictions of the class labels for new orders. A prediction is made by evaluating the attribute vector of a new order.

An SVM requires that each data instance be represented as a vector of real numbers [34]. However, our attributes include categorical features. Hence, before implementation, we need to pre-process data by converting attributes of categorical values into numerical values. We adopt m numbers to represent an $m-$category attribute. A value of 1 means that the instance contains the corresponding category. For instance, the seven-category value-stream attributes $A_5 - A_{11}$ indicate the value streams involved in producing an order. Similarly, for class labels, we use $y_i = 1$ and $y_i = -1$ to represent "On-time order" and "Late order", respectively. This coding technique performs better than the use of a single number to distinguish an attribute of different categories [34].

Another possible conversion of categorical attributes into numerical attributes is to establish a meaningful mapping of each attribute value into a numerical value. For instance, based on manufacturing process complexity of soft-cover types, which is on a scale of (0 to 1), "landscape" corresponds to 0.5 and "minibook" corresponds

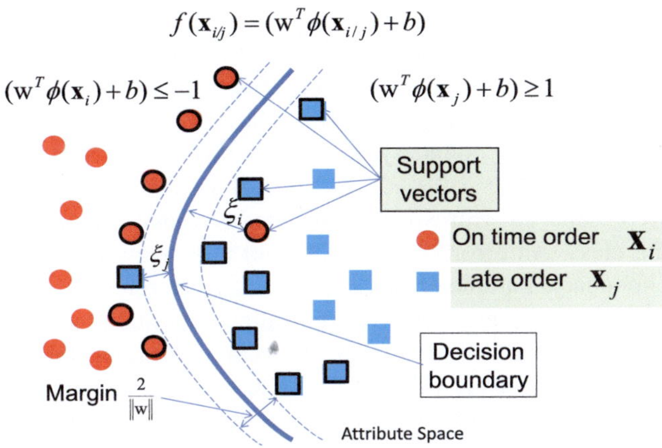

Fig. 5.4 Geometric illustration of the decision boundary in an SVM

to 0.7. This type of conversion requires detailed understanding of the correlation between categorical attributes and manufacturing processes. We do not consider such a targeted mapping technique here since our goal is to study due-date prediction in a general scenario. The proposed method can nevertheless be used with any specific mapping technique in an actual application.

In Fig. 5.4, we illustrate our 14-dimensional problem using a two-dimensional representation. An SVM defines an optimal decision boundary to separate on-time and late orders. The margin from the optimal decision boundary to its closest vectors is the maximum among all possible separating boundaries to their closest vectors. Vectors that are either closest to the decision boundary or on the wrong side of the boundary are called *support vectors*. Only support vectors are maintained by SVM while other vectors are dumped after training since the decision boundary can be uniquely determined by the support vectors. This is an important advantage of SVMs: the classification accuracy does not depend on the size of the training dataset [35].

Our optimization problem can be formulated as follows. We are given a set TD_N, the training dateset of N order/instance-label/class pairs $(\mathbf{x}_i, y_i)$, where $i = 1, \ldots, N$, $\mathbf{x}_i \in R^{14}$, and $y \in \{1, -1\}$. There are N history orders in the dataset, the number of on-time orders is N_1, and the number of late orders is N_2. The SVM computes the unique solution to the following optimization problem [34]:

$$\min_{\mathbf{w},b,\xi} \quad \frac{1}{2}\mathbf{w}^T\mathbf{w} + C_{ontime}\sum_{i=1}^{N_1}\xi_i + C_{late}\sum_{j=N_1+1}^{N}\xi_j \tag{5.3}$$

subject to:

$$y_{i/j}(\mathbf{w}^T\phi(\mathbf{x}_{i/j}) + b) \geq 1 - \xi_{i/j}, \tag{5.4}$$

where $\xi_{i/j} \geq 0$, $i = 1,\ldots,N_1$, $j = N_1 + 1,\ldots,N$. The SVM needs to explore values for vector $\mathbf{w}$, variable b, and vector ξ that minimize the objective function (5.3). According to the properties of an SVM, the distance from the closest vector to the optimal decision boundary is $1/\|\mathbf{w}\|$, where $\mathbf{w}$ is a 14-dimensional vector. The margin of the decision boundary, which is shown in Fig. 5.4, is $2/\|\mathbf{w}\|$. Slack variable $\xi = (\xi_1,\ldots,\xi_i,\ldots,\xi_{N_1},\ldots,\xi_j,\ldots,\xi_N)$ is a N-dimensional vector. Its element ξ_i/ξ_j represents an error when an on-time/late order instance is misclassified into the wrong class. An error is the distance of a wrongly classified instance to the decision boundary. Parameters C_{late} and C_{ontime} are referred to as the *penalty parameters* of the error terms. Since we are dealing with highly imbalanced dataset (more on-time than late orders), penalty parameters for on-time and late orders have different values. The larger the value of C, the heavier the penalty. For instance, if C_{late} is greater than C_{ontime}, a heavier penalty is added for misclassifying a late order as an on-time order than misclassifying an on-time order as a late order.

Furthermore, since orders are inseparable in the linear space defined by the attribute set, the SVM needs to find a linear separating hyperplane (decision boundary plane) with the maximum margin in a higher dimensional space. Instead of showing a straight line in Fig. 5.4, a curve is used to represent the decision boundary projected from the higher dimensional space to the 2-dimensional space. Vectors $\mathbf{x}_{i/j}$ are mapped into a higher dimensional space by the function ϕ. Function $K(\mathbf{x}_i,\mathbf{x}_j) \equiv \phi(\mathbf{x_i})^T\phi(\mathbf{x_j})$ is called the kernel function [35]. We adopt a radial basis kernel function (RBF) defined as follows:

$$K(\mathbf{x}_i,\mathbf{x}_j) = \exp(-\gamma\|\mathbf{x_i} - \mathbf{x_j}\|^2), \quad \gamma > 0 \tag{5.5}$$

where γ is the RBF kernel parameter. The RBF kernel outperforms other kernels such as polynomial and sigmoid kernels in our problem. The term $\mathbf{w}^T\phi(\mathbf{x}_{i/j})$ in Constraint (5.4) is the inner product of $\mathbf{w}$ and $\phi(\mathbf{x}_{i/j})$. Constraint (5.4) is illustrated in Fig. 5.4, which indicates for both classes, the allowed error distance is within $\xi_{i/j}$. The first term in Eq. (5.3) is the inverse of the margin, while the remaining terms equal the sum of errors. The objective function (5.3) is constructed to maximize the margin of the decision boundary while minimizing the sum of classification errors.

When classifying a new order, the SVM computes the value $d = \mathbf{w}^T\phi(\mathbf{x}) + b$ which represents the distance of the new order to the decision boundary. The sign of d determines the class label. The magnitude of $|d|$ determines the value of *Prob* as follows:

$$Prob = \begin{cases} 1 & \text{if } |d| \geq 1, \\ 0.5|d| + 0.5 & \text{otherwise.} \end{cases}$$

In summary, there are three key tuning parameters in our SVM that must be optimized. They are penalty parameters C_{late} and C_{ontime}, and the parameter γ in the RBF kernel. In addition, the popular Sequential Minimal Optimization (SMO) algorithm is implemented to train the SVM model since it is known for its fast convergence [36].

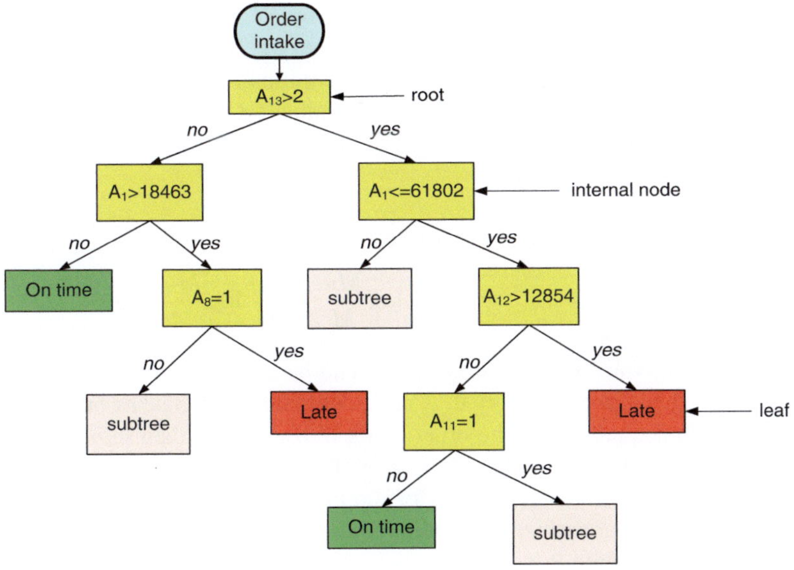

Fig. 5.5 A illustration of a decision tree

5.2.6   Decision Tree

A decision tree learner uses a divide-and-conquer algorithm to construct a tree from a training dataset. It is mostly used to explore features and extract patterns in large databases for classification, regression, and prediction [37]. A decision tree has desirable characteristics and merits such as high accuracy, fast computation, virtually infinite capacity, and intuitive interpretation. Decision trees have been extensively implemented for more than two decades in various areas involving exploratory data analysis and predictive modeling applications [38].

We adopt a nonlinear two-class (binary) decision tree as the second classifier. The training, testing, and validation data sets were real-life data extracted from RPI enterprise resource planning production database for a period of three consecutive weeks. Figure 5.5 shows a simple decision tree. It discovers the underlying rules and regulations to best separate the training and testing datasets into "On-time" and "Late" orders with the least amount of errors. An error refers to a misclassification of an order instance. A brief introduction to decision trees is presented and exemplified by our problem.

A decision tree is a directed acyclic graph consisting of nodes. A decision tree classifier is defined as a classification procedure that recursively partitions a dataset into smaller subdivisions on the basis of a set of rules defined at each node in the tree. In Fig. 5.5, nodes are shown as squares. The tree is composed of a root node (which contains the entire training dataset TD_N), a set of internal nodes, and a set of terminal nodes or leaves [39]. Each node in a binary decision tree has only one parent node and zero or two descendant nodes.

Every node contains a set of orders denoted by STD_M, where M is the size of STD_M and $STD_M \subseteq TD_N$ (i.e., STD_M is a subset of TD_N, $M \leq N$; for the root node, $STD_M = TD_N$). In each dataset STD_M, the number of late and on-time orders is denoted by M_P and M_N, respectively. An evaluation matrix called *information value* or *information entropy* [40], which is denoted by parameter *info*, is defined and evaluated at each node as follows:

$$info(STD_M) = info([M_P, M_N]) \tag{5.6}$$

$$= -\frac{M_P}{M} \log \frac{M_P}{M} - \frac{M_N}{M} \log \frac{M_N}{M} \tag{5.7}$$

where $M_P + M_N = M$. The property of *info* can be expressed as follows. If M_P or M_N equal to 0, then $info(STD_M) = 0$. A node containing only late or on-time orders has an *info* value equals 0. On the other hand, if $M_P = M_N$, then $info(STD_M) = 1$ and it reaches its maximum value. A node with $info = 1$ has the same number of late and on-time orders. Hence, regularity and unambiguity of the order completion status at a node are characterized by small *info*, whereas, randomness and ambiguity are characterized by large *info*.

We adopt the popular greedy decision-tree algorithm "maximize the information gain" to construct the decision tree [37]. Starting from the root node, if a node has nonzero *info*, then its dataset STD_M should be further divided into two non-overlapping subsets STD_{M1} and STD_{M2} by a set of rules. These rules divide the orders by checking their attribute values. Two descendent nodes are created and each of them contains STD_{M1} and STD_{M2}, respectively. The algorithm requires that after the division of STD_M at a node, the information value gained from this separation should be maximized. The objective is formulated as follows:

$$\text{maximize } info(STD_M) - (info(STD_{M_1}) + info(STD_{M_2})) \tag{5.8}$$

A decision tree is constructed by exploring the rules that maximize the above objective at each node until all descendant nodes are leaves. There are three conditions under which a node becomes a leaf. (i) If its *info* is zero, i.e., a node contains only one type of orders. (ii) Its *info* is nonzero, however, orders in its dataset are indistinguishable by any attribute except for their class labels, i.e., its dataset cannot be separated by any rule. (iii) Its *info* is nonzero, the ratio α, which is defined as the ratio of the number of "Late" orders versus "On-time" orders, is below a certain threshold α_{th}. Parameter α_{th} is used to limit the depth of a decision tree.

Once constructed from a training dataset, a decision tree can be used for classification. A new order instance represented by input vector $\mathbf{x_j}$ traverses the decision tree from the root node to a leaf. The new order $\mathbf{x_j}$ is classified as the dominant order type at the leaf. The probability estimation *Prob* is the ratio of the number of dominant orders versus the number of all orders at the leaf. A leaf represents a classification decision. In our problem, a leaf is either an "On-time order" or a "Late order" class label. Every path from the root node to a leaf can be expressed as a decision rule, which is a combination of rules at the nodes along the path. An example decision rule is as follows:

$$\text{if } A_2 \text{ is Cover Only and } A_4 > 1 \text{ day , then } y = \text{"On-time order"}$$

where A_2 and A_4 are the second and the fourth attribute in the input vector, respectively. This decision rule corresponds to the left-most path in the decision tree shown in Fig. 5.5.

In the real-life case studies that we considered, the training dataset contained more than 4,000 order instances. Although only 14 attributes were used, the final decision tree involves 117 internal nodes and more than 59 leaves. It is difficult to present the entire decision tree in one picture. Therefore, part of the decision tree is illustrated in Fig. 5.5.

5.2.7 Bayesian Probabilistic Model

The third classifier that we adopt is an efficient probabilistic model based on Bayes' theorem [41]. An illustration of this model is shown in Fig. 5.6. By counting the number of on-time and late history orders that have been completed, we compute two prior probabilities $P(OnTime)$ and $P(Late)$, where $OnTime$ and $Late$ are the events that an order is delivered on time or late, respectively. Probability $P(OnTime)$ is approximated by the ratio of the number of on-time orders versus all orders. Probability $P(Late)$ equals $1 - P(OnTime)$. Two likelihoods for every value a of every attribute A of history orders, $P(A = a|OnTime)$ and $P(A = a|Late)$, are also computed by calculating the occurrence frequencies of $A = a$ for on-time and late orders, respectively. All these prior probabilities and likelihoods are stored in the knowledge base and updated whenever there are orders shipped on time or late. Therefore, unlike an SVM or a decision tree, the quantity of training orders is important in a probabilistic model. The more the orders, the closer the occurrence frequency approximates the actual probability distribution.

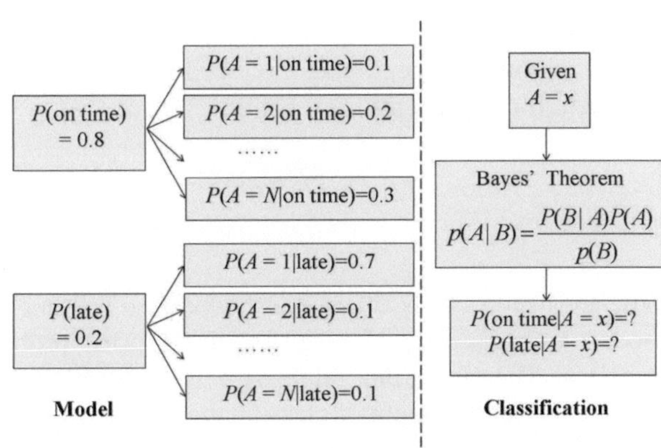

Fig. 5.6 An illustration of the Bayesian probabilistic model

When given a new input vector $\mathbf{x_j}$ for prediction, the probabilistic model first applies Bayes' theorem to obtain the posterior probabilities conditioned on each attribute. Given an attribute $A = a'$ in $\mathbf{x_j}$, its two posterior probabilities are calculated as follows:

$$P(OnTime|A = a') = \frac{P(A = a'|OnTime) \cdot P(OnTime)}{P(A = a')} \tag{5.9}$$

$$P(Late|A = a') = \frac{P(A = a'|Late) \cdot P(Late)}{P(A = a')} \tag{5.10}$$

where $P(OnTime|A = a')$ and $P(Late|A = a')$ are the probabilities that this contract $\mathbf{x_j}$ is going to be shipped on time or late conditioned on its attribute A equals a', respectively. The common term $P(A = a')$ in the denominators is set to 1.

We also assume that all the attributes A_1, A_2, $\ldots$, A_{14} are independent of each other. This assumption does not lose generality since orders are very diverse in terms of attributes, and the quantity of past orders is considerable. Furthermore, no significant correlation was found among any two attributes in our evaluation. The property of independent events states that events B and C are said to be independent if [42]:

$$P(B \cap C) = P(B) \cdot P(C) \tag{5.11}$$

Therefore, by applying the property of independent events, the probability that order $\mathbf{x_j}$ will be shipped on time or late while considering all the attributes is computed as follows:

$$P(OnTime|\mathbf{x_j}) = P(OnTime|A_1 = a_1) \cdot \ldots \cdot P(OnTime|A_{14} = a_{14}) \tag{5.12}$$

$$P(Late|\mathbf{x_j}) = P(Late|A_1 = a_1) \cdot \ldots \cdot P(Late|A_{14} = a_{14}) \tag{5.13}$$

where each term on the right hand side is calculated by Eqs. (5.9) or (5.10).

In addition, we define a parameter δ to enhance the adaptability and controllability of the probabilistic model. As mentioned before, the numbers of late and on-time orders are heavily imbalanced since there were more on-time orders than late ones. The due-date validation engine will further amplify the imbalance. If decisions given by the engine were deployed, more orders could be delivered on time. Parameter δ is used to modify the prior probabilities, such that,

$$P'(Late) = P(Late) - \delta \tag{5.14}$$

$$P'(OnTime) = P(OnTime) + \delta \tag{5.15}$$

Whenever an order is shipped, it becomes a training instance for the probabilistic model. Consequently, the values of all prior probabilities and likelihoods are transient. Therefore, the value of δ is also trained and updated dynamically to achieve the least amount of classification errors.

In the final step of prediction, since probabilities $P'(OnTime|\mathbf{x_j})$ and $P'(Late|\mathbf{x_j})$ are calculated, the probabilistic model sets the predicted class label for order $\mathbf{x_j}$ to be the one that has a higher value. Therefore the probability estimation *Prob* is the corresponding P'.

5.2.8   *Comparison of Classifiers*

This subsection summarizes similarities and differences between the three machine-learning classifiers. First, an SVM is an instance-based classifier whose decision boundary is defined by representative instances known as support vectors. A decision tree is a rule-based classifier and it is equivalent to a set of decision rules. A probabilistic model is a probability-based classifier, whose final result is obtained by comparing probabilities. Second, an SVM is a nonlinear classifier in the attribute space. It finds the decision boundary by transforming the attribute space into a higher-dimensional space with the application of a kernel function. A decision tree is a nonlinear classifier in the attribute space [39]. A probabilistic model can be viewed as a linear classifier. By taking logs of both sides of Eqs. (5.12) and (5.13), we obtain the following transformations:

$$log(P(OnTime|\mathbf{x_j})) = \sum_{i=1}^{14} log(P(OnTime|A_i = a_i)) \qquad (5.16)$$

$$log(P(Late|\mathbf{x_j})) = \sum_{i=1}^{14} log(P(Late|A_i = a_i)) \qquad (5.17)$$

Therefore, the value of $log(P(OnTime|\mathbf{x_j}))$ or $log(P(Late|\mathbf{x_j}))$ is a linear combination of $log(P(OnTime|A_i = a_i))$ or $log(P(Late|A_i = a_i))$, where $1 \leq i \leq 14$. Furthermore, the performance of an SVM or a decision tree depends on the representative order instances rather than the quantity of instances. A probabilistic model requires a sufficient number of order instances to compute prior probabilities and likelihoods [43]. Since the classifiers are diverse as discussed above, we consider a combination of them to achieve better prediction.

In addition, since our problem is linearly inseparable, the machine-learning algorithms we selected do not estimate weights for each attribute. An SVM maintains support vectors instead of evaluating each attribute. A decision tree is a nonlinear classifier that does not evaluate weights for each attribute. The Bayes probabilistic model was built under the assumption that each attribute is independent and equally important.

5.3   Decision Integration

In this section, we describe decision integration strategies for increasing the accuracy of classification. Inputs to the decision integration center are the classification results (class labels and probability estimations) given by each classifier. Each classifier has its own advantages, constraints and applicability determined by their underlying algorithms. Therefore, the predictions for the same order given by the three classifiers may be different. Decision integration is implemented in the interest of preserving the advantage of each classifier. There are some popular candidates such as majority voting, weighted voting, and so on [44]. We considered three decision integration approaches: Dempster-Shafer theory based decision integration approach, decision fusion approach, and voting approach. The decision integration center can select an approach for a given target scenario.

5.3.1   Dempster-Shafer Theory-Based Decision Integration Approach

Dempster-Shafer theory (DST) is a mathematical theory of evidence that allows one to combine evidence from different sources and arrive at a degree of belief that takes into account all the available evidence [45]. It is often used to deal with problems that are lack of parametric probability distribution models or accurate mathematical analysis [46]. The implementation of DST in our problem is presented below.

In a Dempster-Shafer reasoning system, all possible mutually exclusive facts are enumerated in "the frame of discernment Θ" [47]. In our problem, an order is either a "Late" or an "On-time" order, which are represented as facts P (positive) and N (negative), respectively. Therefore, in our problem, $\Theta = \{P, N\}$ to indicate whether an order is late or on time.

Classifier S_i contributes its prediction for an order by assigning its beliefs over Θ. This assignment function is called the "probability mass function" of classifier S_i, and it is denoted by m_i. According to classifier S_i's prediction, the probability that "the order is late" is indicated by a "confidence interval" [$\text{Belief}_i(P)$, $\text{Plausibility}_i(P)$]. For every classifier, there are only four possible evidences for an order and they are described in Table 5.2, in which TP, TN, FP, and FN rates are classifier performance metrics (Sect. 5.2.4), and parameters $Prob(P)$ and $Prob(N)$ are classifier outputs (Sect. 5.2.3). Evidence E_1 and E_2 support proposition P ($A = \{E_1, E_2\}$), whereas evidence E_3 and E_4 support proposition N.

The lower bound of the confidence interval is the belief confidence, which accounts for all evidence E_k that supports the given proposition P. Therefore:

$$\text{Belief}_i(P) = \sum_{E_1, E_2 \subseteq A} m_i(E_k) = m_i(E_1) + m_i(E_2) \qquad (5.18)$$

Table 5.2 Four possible evidences of the completion status for an order

Evidence	Real order status	Predicted order status	$m(E_k)$
E_1	P	P	$TP \cdot Prob(P)$
E_2	P	N	$FN \cdot Prob(N)$
E_3	N	N	$TN \cdot Prob(N)$
E_4	N	P	$FP \cdot Prob(P)$

The upper bound of the confidence interval is the plausibility confidence, which accounts for all the evidence that do not rule out the given proposition P. Therefore,

$$\text{Plausibility}_i(P) = 1 - \sum_{E_3, E_4 \bigcap A = \emptyset} m_i(E_k) = 1 - m_i(E_3) - m_i(E_4) \qquad (5.19)$$

For each possible proposition, e.g., P, DST gives a rule of combining classifier S_i's prediction m_i and classifier S_j's prediction m_j as follows:

$$(m_i \bigoplus m_j)(P) = \frac{\sum_{E_k \bigcap E_{k'} \subset A} m_i(E_k) m_j(E_{k'})}{1 - \sum_{E_k \bigcap E_{k'} = \emptyset} m_i(E_k) m_j(E_{k'})}. \qquad (5.20)$$

Furthermore, this combination rule can be generalized using iteration. After combining the predictions of two classifiers, the combined result can be treated as a new classifier such that it can be further combined with the third classifier.

5.3.2 Decision Fusion Approach

In this subsection, we introduce the decision fusion approach, demonstrate the implementation in our problem, and discuss its properties.

Decision fusion techniques have been applied to areas such as computer-aided diagnostic models for breast cancer [48], distributed sensor surveillance systems [49], and other fields. In these fields, for instance, medical imaging and distrusted sensing, the sensing events are immersed in significant noise from diverse sources. Decision fusion techniques are used to primarily detect if the true signature is present.

5.3.2.1 Implementation

The due-date validation engine also uses decision fusion technique as one integration framework for multiple classifiers. In our problem, late deliveries occur less frequently than on-time deliveries and incur extra costs for the factory. Late deliveries should be captured more accurately than on-time deliveries, therefore, they are referred to as positive signals that need to be detected.

Each classifier makes either a "positive" (late order) or a "negative" (on-time order) prediction for a new order. Note that, alternatively, on-time orders can be treated as positive signals. As long as classifiers are the same, the integration results are not affected by taking which class label as the positive signal.

There are four conditional probabilities denoted as $P(p|p)$, $P(n|p)$, $P(p|n)$, and, $P(n|n)$, which represent the probability of classifying an order as positive/negative conditioned on its actual completion-status as positive/negative. For example, parameter $P(p|p)$ is the probability of correctly classifying a late order. Therefore, $P(p|p)$ can be estimated by the TP rate.

For the ith classifier, variable d_i denotes its prediction decision for an order, where $d_i \in \{p, n\}$. We define a parameter $\lambda(d_i) = P(d_i|p)/P(d_i|n)$. Therefore, if $d_i = p$, then $\lambda(d_i) = P_i(p|p)/P_i(p|n) = TP_i/FP_i$. Similarly, if $d_i = n$, then $\lambda(d_i) = P_i(n|p)/P_i(n|n) = FN_i/TN_i$. Note that, since each classifier makes a binary decision, there are only two possible values of $\lambda(d_i)$, i.e., $\lambda(d_i = p)$ and $\lambda(d_i = n)$.

In decision fusion approach, positive signal $\lambda(d_i = p)$ should always be greater than negative signal $\lambda(d_i = n)$. Otherwise, the value of $\lambda(d_i)$ is in contradiction to signal definition. We then have the following:

$$\lambda(d_i = p) > \lambda(d_i = n)$$

$$\frac{TP}{FP} > \frac{1 - TP}{1 - FP}$$

$$\Rightarrow TP > FP$$

Therefore, in order for a classifier to be used in decision fusion approach, its TP must by greater than FP.

The prediction of each classifier is independent. Parameter $\lambda(d)$ is defined as $\lambda(d) = \lambda(d_i) \cdot \lambda(d_2) \cdots \lambda(d_N)$. Therefore, each new order has its value λ computed based on predictions from classifiers as follows:

$$\lambda(d) = \frac{\prod_{i=1}^{N} p(d_i|p)}{\prod_{i=1}^{N} p(d_i|n)} \tag{5.21}$$

Note that, since there are only two possible values of $\lambda(d_i)$, the total possible values of λ for a new order is 2^N. They can be referred to as λ_1 to λ_{2^N} in ascending order.

We then define parameter λ_{th} as a threshold value which is used to determine the final prediction. If $\lambda < \lambda_{th}$, then the final prediction is negative. If $\lambda > \lambda_{th}$, then the final prediction is positive.

As shown in Fig. 5.7, the number of possible positions for λ_{th} when $N = 3$ is $2^3 - 1 = 7$. Note that, it is useless for λ_{th} to be less than λ_1 or greater than λ_{2^N} since otherwise the final prediction is uniformly negative or positive. We compute $\lambda_{thj} = (\lambda_j + \lambda_{j+1})/2$.

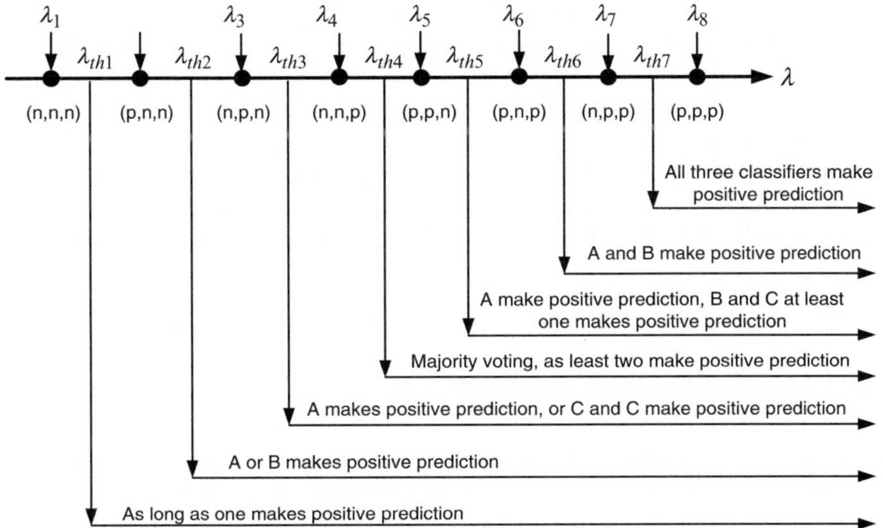

Fig. 5.7 Demonstration of the equivalence of the decision fusion approach to multiple voting strategies. Each tuple (n/p, n/p, n/p) represents the prediction results of classifier (A, B, C)

Therefore, based on each threshold value, the three classifiers generate a new classifier. Its performance can be evaluated by the same metrics as for other classifiers.

5.3.2.2   Properties

In this subsection, we demonstrate that the decision fusion approach covers multiple voting strategies that can be used for integrating multiple decisions. First, we rank the three classifiers by $\lambda(d_i = p)/\lambda(d_i = n)$ in descending order and rename them as classifier A, B, and C. As shown in Fig. 5.7, each tuple (n/p, n/p, n/p) represents the prediction results of classifier (A, B, C). Furthermore, each tuple corresponds to a unique value of λ, which is shown in Fig. 5.7. As we have discussed, the seven threshold values λ_{th} correspond to seven classifiers, and they further correspond to seven voting strategies as demonstrated in Fig. 5.7. For example, only a tuple with at least two positive predictions has a λ greater than λ_{th4}. Therefore, parameter λ_{th4} corresponds to majority voting strategy.

5.3.3   Voting Approach

The voting approach deploys a simple and efficient joint method to integrate three class labels and it determines that the final class label is "Late order" only when all the three input class labels are "Late order". If any classifier inputs an "On-time order" label, the final class label is "On-time order".

The reason for deploying this voting strategy is because *FP* rates of all classifiers are high. This kind of strategy takes the advantage of the differences among classifiers. Only on-time orders that they all misclassified as late orders are the final misclassified orders that contribute to the final *FP* rate. The smaller the overlap, the lower the final *FP* rate.

Similar to the DST-based approach, the voting approach also generates one new classifier. Therefore, when we apply three decision integration approaches and use three classifiers, there can be a total $3 + 1 + 7 + 1 = 12$ classifiers that can run in parallel.

5.3.4 Exploring New Due Dates

After a new order is predicted to be late, the next step is to explore the appropriate due date. If we continue increasing the due date, there is a value that changes the predicted order-completion status from late to on-time. This value is illustrated in Fig. 5.8 and denoted as *dp*, where parameter *DL* and *DU* denote the lowest and highest allocated manufacturing time for orders that have ever seen in history. Therefore, we define a "phase-change point" for due date to be a value that flips the predicted order-completion status from "late" to "on-time". In order to enhance the efficiency of reaching to the phase-change point, we adopted a binary search method and ten steps showed good convergence performance for real-life order data.

In comparison with the proposed classifier-based due-date exploration method, conventional due-date prediction methods build a regression model. The application of SVM for regression model is called support vector regression (SVR). SVR has been successfully applied in various fields, e.g., time series and financial prediction [50], forecasting of electricity prices [51], and estimation of power consumption [52]. A tutorial in [53] has discussed the process of selecting algorithms for training SVR models. Since there are many successful applications with SVR models, it motivates our research in using SVR as the baseline model.

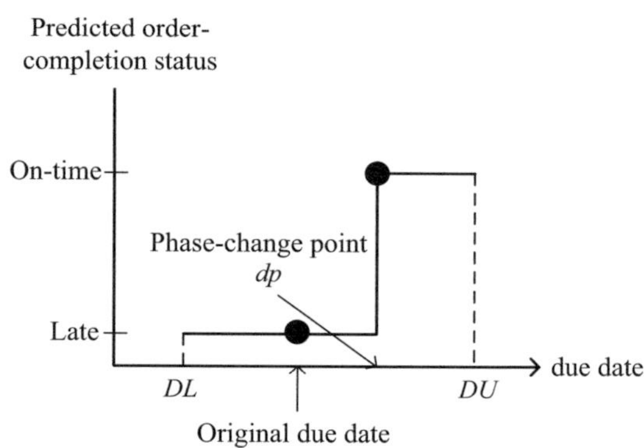

Fig. 5.8 Illustration of a phase-change point for a numerical attribute

The SVR model can be easily used to predict order completion status. By comparing the predicted due date with the proposed due date, the SVR model classifies an order as an "On-time order" if the predicted due date is no greater than the proposed due date and as a "Late order" otherwise.

5.4 Results and Discussions

In this section, we evaluate the proposed due-date validation engine using real-life order data from RPI. Our goal is to show that we can accurately predict which orders will miss their designated due dates. Such prediction results can be used to negotiate new due dates with customers.

The SVM, SVR, and decision tree model were implemented using the open-source machine-learning Weka software [22]. Original order due dates were assigned by the template-based method as discussed in Sect. 5.1.1.

5.4.1 Classifier Evaluation Strategy and Results

The training, testing, and validation data sets were three consecutive weeks real-life data extracted from RPI enterprise resource planning database. This data therefore corresponds to a real production environment. The first, second, and third week's data were used as training, testing, and validation sets, respectively, as shown in Fig. 5.9. Parameters of the three classifiers and the SVR model were optimized such that each classifier can reach the TP rate no less than 0.95 and a FP rate as low as possible. Classifier prediction performance was evaluated by order instances in the validation set.

The comparison results for different classifiers and the SVR model with and without using factory-property attributes and evaluation metrics are shown in Table 5.3, Figs. 5.10 and 5.11.

Table 5.3 also summarizes the parameters that need to be optimized for each classifier. They are the penalty parameters C_{late} and C_{ontime} shown in Eq. (5.3), parameter γ in a RBF kernel shown in Eq. (5.5), parameter α_{th} used for constructing

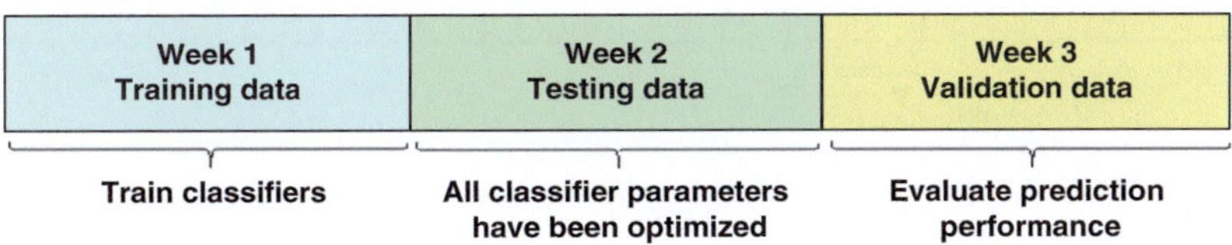

Fig. 5.9 Partitioning of data for training, testing, and validation

Table 5.3 Comparison results and tunable parameters for classifiers

Classifier	With	Without	With	Without	Tunable parameters
	Factory-property attributes				
	TP rate		*FP* rate		
SVM	0.973	0.987	0.506	0.804	$C_{late}, C_{ontime}, \gamma$
Decision tree	0.990	0.993	0.713	0.887	α_{th}
Bayesian probabilistic model	0.994	0.997	0.512	0.726	δ
Dempster-Shafer theory	0.966	0.980	0.173	0.545	None
λ_{th1}	0.999	1.000	0.765	0.923	
λ_{th2}	0.994	0.999	0.649	0.774	
λ_{th3}	0.990	0.998	0.589	0.767	
λ_{th4}	0.988	0.996	0.526	0.652	λ_{th}
λ_{th5}	0.977	0.982	0.323	0.622	
λ_{th6}	0.969	0.977	0.160	0.549	
λ_{th7}	0.946	0.963	0.145	0.533	
Voting approach	0.949	0.970	0.149	0.532	None
SVR	0.975	0.988	0.623	0.797	$C_{late}, C_{ontime}, \gamma$
Decision integration center	None				Integration approaches

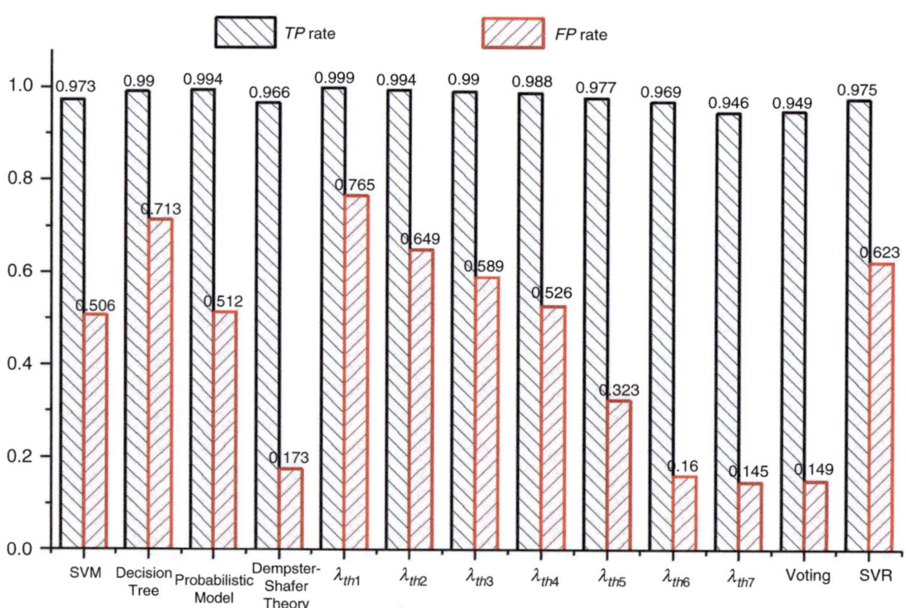

Fig. 5.10 Comparison results of *TP* and *FP* rates of classifiers and the SVR model with factory-property attributes

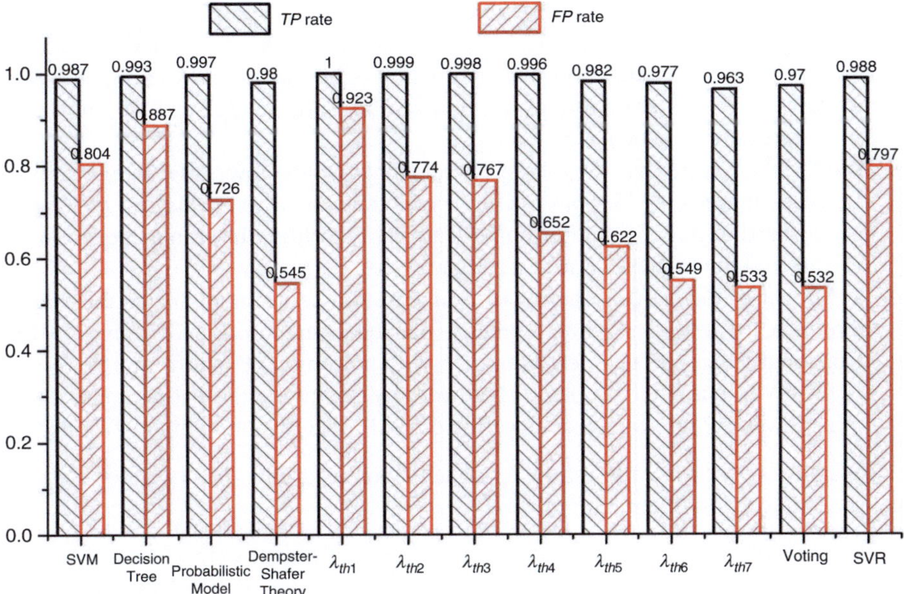

Fig. 5.11 Comparison results of *TP* and *FP* rates of classifiers and the SVR model without factory-property attributes

a decision tree, parameter δ shown in Eq. (5.14), and the most efficient decision integration approach among the three options we have evaluated. If the decision-fusion approach is adopted, then the value of λ_{th} should be optimized.

5.4.2   Discussion

The most significant computation-time component for the proposed framework is model development for machine learning; however, this computation is done offline and it can be parallelized. The model prediction and decision integration take less than a second of computation time. Once these models are developed, they can be deployed and used in a real-time environment. The software development is indeed feasible for a print factory. The machine learning is done using open-source software, hence there is no need for a significant amount of in-house software development.

The following key conclusions can be drawn from the experimental results:

1. Factory-property attribute data constitute important information for predicting order due date and order completion status. While maintaining the same level of *TP* rates, the addition of factory-property attributes can significantly reduce the *FP* rates than just using order-property attribute data.

2. Each individual classifier offers a trade-off between a high *TP* rate, i.e., "Late order" prediction accuracy, and a low *FP* rate, i.e., "On-time order" misclassification probability. However, the optimal classifier should have a high *TP* rate and a low *FP* rate. Even though *TP* rate can be made as high as 0.96, the value of *FP* rate is no less than 0.5.
3. The SVM classifier has the lowest *FP* rate and also the lowest *TP* rate, whereas, the decision tree classifier has the highest *TP* rate and also the highest *FP* rate. It is very hard to find a single machine-learning algorithm that can achieve lowest *FP* rate and the highest *TP* rate simultaneously.
4. As shown in Fig. 5.10, if the decision integration center implements the voting approach, the value of *FP* rate is greatly reduced from more than 0.7 (given by the decision tree classifier) to less than 0.15 with a drop of *TP* rate from 0.96 (given by the SVM classifier) to 0.94. Therefore, by deploying the voting approach, a much lower *FP* rate can be achieved with negligible reduction in the *TP* rate.

In addition, we draw the following conclusions regarding the choice of classification and decision integration methods.

1. The proposed multi-classifer decision-integration model outperforms the SVR model. Classifiers have the advantages of generating probabilistic estimates, which regression models do not have. Predicted probabilities from different classifiers form another parameter for the decision-integration center to tune the performance of the integrated model. Such additional parameters enhance the flexibility and prediction accuracy of the integrated model.
2. The performance of the DST-based approach is similar to classifications resulting from the decision fusion approach. For the decision fusion approach, when λ_{th7} is used, the performance is the same as the voting approach. This can be explained by the properties of the decision fusion approach discussed in Sect. 3.2.2.
3. By comparing Fig. 5.7 and Table 5.3, we can seen that classifiers correspond to λ_{th1}, λ_{th2}, and λ_{th3} can still increase *TP* rate compared to individual classifiers. Starting from λ_{th4}, the *TP* rate becomes lower than individual classifiers. This is because the voting strategy of an order being classified as a late order becomes stricter as λ_{th} increases. As demonstrated by Fig. 5.7, fewer orders can be classified as late orders as λ_{th} increases. Although some late orders have been correctly identified by individual classifiers, the voting strategy rules them out.
4. Different λ_{th} value results in different classifier performance. Depending on the requirements of *TP* or *TN* rates, the decision fusion approach provides more choices for optimizing the integrated results compared to other approaches. Demands on *TP* and *TN* rates vary with the factory needs. For instance, if late-delivery associated cost is high, then a high *TP* rate is desired. If postponing due dates cost more, then a low *FP* rate is most demanding. Decision fusion approach provides more options on *TP* and *TN* rates. The threshold value can be adjusted according to the actual need.

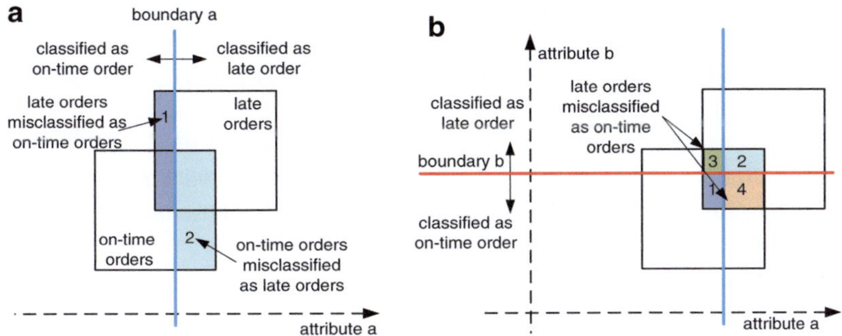

Fig. 5.12 An example for the order completion-status boundary in (**a**) one dimension; (**b**) two dimensions

In summary, the problem we have attempted to solve is to enhance the accuracy of identifying orders that are close to the order completion-status boundary. The main difficulty in prediction is caused by overlapping of on-time and late orders in the order-attribute space. The boundary can be placed in favor of decreasing *FN* rate with the inevitable increase in *FP* rate, or vice versa. For instance, we can aim for much higher *FP* rates than *FN* rates of all three classifiers.

Even if the boundary can be well-defined, as illustrated in Fig. 5.12a, the lowest *FP* rate that the static template-based method can achieve is as good as an individual classifier (e.g., 0.508), since the static method cannot distinguish orders within the overlap area. By adopting decision integration, the data-driven approach significantly reduces the *FP* rate (e.g., to only 0.142). Based on history and up-to-date orders, the data-driven approach tackles this challenge by estimating the probability that an order is placed on either side of and close to the boundary.

Decision integration provides more options for making the final prediction. It is equivalent to adding an extra attribute to the attribute space. Compared to Fig. 5.12a, an extra attribute b is added by adopting decision integration; therefore, boundary b can be established. The overlap area is divided into four regions by the two boundaries. Region 1 contains late orders that are misclassified as on-time orders. Region 2 contains on-time orders that are misclassified as late orders. Region 3 and Region 4 have conflict classifications given by the two boundaries. One possible integration strategy can be defined such that orders in both Region 3 and Region 4 are all classified as on-time orders.

We then evaluate *FN* and *FP* rates in both cases. Areas of late and on-time orders are the same and denoted by area{late orders} and area{on-time orders}, respectively. Let *FN*(*a*) denote the false negative rate of classification in case (a). Hence, *FN*(*a*) = area{region 1}/area{late orders}= 6/36, *FP*(*a*) = area{region 2}/{on-time orders}= 12/36. Similarly, let *FN*(*b*) denote the false negative rate of classification in case (b). Hence, $FN(b) = \text{area}\{\text{region } 1, 3, 4\}/\text{area}\{\text{late orders}\} = 7/36$, $FP(a) = \text{area}\{\text{region } 2\}/\{\text{on-time orders}\} = 2/36$. We can see that according to the decision integration strategy, FP rate is reduced from 12/26 to 2/36, whereas FN rate is only increased from 6/36 to 7/36.

Therefore, as the complexity and diversity of both orders and the factory increase, a static prediction method becomes even more inaccurate, and a data-driven order-admission policy, as presented in this chapter, must be adopted to optimize performance.

5.5   Conclusion

We have presented an intelligent order-admission framework that predicts order completion status and due dates in real time for each incoming order using machine-learning and decision-integration techniques. We have considered the impact of diverse order-property attributes and the importance of dynamic factory-property attributes. We have deployed three powerful machine-learning algorithms in the framework: support vector machine, decision tree, and Bayesian probabilistic model. The classifiers have been trained and optimized using history orders. Three decision integration approaches have been applied to derive a final prediction. The performance of classifiers has been tested using real-order data from a PSP. Results show that the integration strategy outperforms stand-alone classification methods as well as the support vector regression model.

References

1. R. Rosales, M. Glass, J. Teich, B. Wang, Y. Xu, R. Hasholzner, M aestroholistic actor-oriented modeling of nonfunctional properties and firmware behavior for MPSoCs. ACM Trans. Des. Autom. Electron. Syst. (TODAES) **19**(3), 23 (2014)
2. C.-H. Tu, H.-H. Hsu, J.-H. Chen, C.-H. Chen, S.-H. Hung, Performance and power profiling for emulated android systems. ACM Trans. Des. Autom. Electron. Syst. (TODAES) **19**(2), 10 (2014)
3. B.H. Meyer, A.S. Hartman, D.E. Thomas, Cost-effective lifetime and yield optimization for NoC-based MPSoCs. ACM Trans. Des. Autom. Electron. Syst. (TODAES) **19**(2), 12 (2014)
4. R. Schneider, D. Goswami, S. Chakraborty, U. Bordoloi, P. Eles, Z. Peng, Quantifying notions of extensibility in flexray schedule synthesis. ACM Trans. Des. Autom. Electron. Syst. (TODAES) **19**(4), 32 (2014)
5. H. Ding, Y. Liang, T. Mitra, WCET-centric partial instruction cache locking, in *2012 49th ACM/EDAC/IEEE Design Automation Conference (DAC)*, San Francisco, June 2012, pp. 412–420
6. E. Ma, R. Yan, K. Huang, M. Yu, S. Xiu, H. Ge, X. Yan, A. Jerraya, Performance estimation techniques with MPSoC transaction-accurate models. IEEE Trans. Comput. Aided Des. Integr. Circuits Syst. **32**(12), 1920–1933 (2013)
7. J. Zeng, I.-J. Lin, E. Hoarau, G. Dispoto, Next-generation commercial print infrastructure: Gutenberg-Landa TCP/IP as cyber-physical system. J. Imaging Sci. Technol. **54**(1), 1–6 (2010)

8. J. Zeng, S. Jackson, I. Lin, M. Gustafson, E. Gustafson, R. Mitchell, Operations simulation of on-demand digital print, in *IEEE 13th International Conference on Computer Science and Information Technology*, Chengdu (2011)
9. J. Spohrer, P.P. Maglio, J. Bailey, D. Gruhl, Steps toward a science of service systems. IEEE Comput. Soc. **40**(1), 71–77 (2007)
10. J. Zeng, I.-J. Lin, G. Dispoto, E. Hoarau, G. Beretta, On-demand digital print services: a new commercial print paradigm as an it service vertical, in *Annual SRII Global Conference*, Silicon Valley (2011), pp. 120–125
11. S. Karp, The future of print publishing and paid content, Publishing 2.0 blog. Tech. Rep. (2007), http://publishing2.com/2007/12/06/the-future-of-print-publishing-and-paid-content/
12. H. Kipphan, *Handbook of Print Media: Technologies and Production Methods* (Springer, New York, 2001), no. 40–422
13. L.N. Nassif, J.M. Nogueira, A. Karmouch, M. Ahmed, F.V.D. Andrade, Job completion prediction using case-based reasoning for grid computing environments. Concurr. Comput. Pract. Exp. **19**, 1253–1269 (2007)
14. R. Pongracz, J. Bartholy, I. Bogardi, Fuzzy rule-based prediction of monthly precipitation. Phys. Chem. Earth **9**, 663–667 (2001)
15. S.F. Smith, Knowledge-based production management: approaches, results and prospects. Prod. Plann. Control **3**, 350–380 (1993)
16. L. Mandrake, U. Rebbapragada, K.L. Wagstaff, D. Thompson, S. Chien et al., Surface sulfur detection via remote sensing and onboard classification. ACM Trans. Intell. Syst. Technol. **3**(4), 77 (2012)
17. K. Chung, C.P. Sang, A machine learning approach to yield management in semiconductor manufacturing. Int. J. Prod. Res. **38**(17), 4261–4271 (2000)
18. X.S. Zhang, B. Shrestha, S. Yoon, S. Kambhampati, P. DiBona et al., An ensemble architecture for learning complex problem-solving techniques from demonstration. Int. J. Prod. Res. **4**(3), 75 (2012)
19. L. Monostori, AI and machine learning techniques for managing complexity, changes and uncertainties in manufacturing. Eng. Appl. AI **16**(4), 277–291 (2003)
20. P. Kaminsky, D. Hochbaum, Due date quotation models and algorithms, in *Handbook of Scheduling: Algorithms, Models, and Performance Analysis* (CRC Press, Boca Raton, 2004), pp. 20–21
21. Q. Duan, J. Zeng, K. Chakrabarty, G. Dispoto, Real-time production scheduler for digital-print-service providers based on a dynamic incremental evolutionary algorithm. Accepted for publication in IEEE Trans. Autom. Sci. Eng. Available on IEEE Xplore as Early Access article, vol. PP, no. 99, pp. 1–15 (2014)
22. M. Hall, E. Frank, G. Holmes, B. Pfahringer, P. Reutemann, I.H. Witten, The weka data mining software: an update. SIGKDD Explor. **11**(1), (2009). Available: http://www.cs.waikato.ac.nz/ml/weka/
23. Y. Mao, W. Chen, Y. Chen, C. Lu, M. Kollef, T. Bailey, An integrated data mining approach to real-time clinical monitoring and deterioration warning, in *Proceedings of the ACM SIGKDD Conference*, Beijing, vol. 12 (2012)
24. X. Zhao, Y. Chen, W. Zhang, An efficient and integrated strategy for temporal planning, in *The Third International Conference on Integration of AI and OR Techniques in Constraint Programming for Combinatorial Optimization Problems*, Cork (2006), pp. 273–287
25. J. Han, K. Micheline, *Data Mining: Concepts and Techniques*, vol. 5 (Morgan Kaufmann, San Francisco, 2001)
26. I. Guyon, A. Elisseeff, An introduction to variable and feature selection. J. Mach. Learn. Res. **3**, 1157–1182 (2003)
27. X. Wu, V. Kumar, J.R. Quinlan, J. Ghosh, Q. Yang, H. Motoda, G.J. McLachlan, A. Ng, B. Liu, S.Y. Philip et al., Top 10 algorithms in data mining. Knowl. Inf. Syst. **14**(1), 1–37 (2008)
28. J. Huang, J. Lu, C.X. Ling, Comparing naive bayes, decision trees, and SVM with AUC and accuracy, in *The Third IEEE International Conference on Data Mining*, Melbourne (2003), pp. 553–556

29. D.M.W. Powers, Evaluation: from precision, recall and f-factor to ROC, informedness, markedness & correlation. J. Mach. Learn. Technol. **2**(1), 37–63 (2011)
30. V.N. Vapnik, *The Nature of Statistical Learning Theory* (Springer, New York, 1995), no. 40–422
31. B.L. Milenova, J.S. Yarmus, M.M. Campos, SVM in oracle database 10g: removing the barriers to widespread adoption of support vector machines, in *Proceedings of the 31st VLDB Conference*, Trondheim (2005)
32. O. Chapelle, P. Halffner, V. Vapnik, Support vector machines for histogram-based image classification. IEEE Trans. Neural Netw. **10**(5), 1055–1064 (1999)
33. G. Gao, Y. Zhu, G. Duan, Y. Zhang, Intelligent fault identification based on wavelet packet energy analysis and SVM, in *International Conference on Control, Automation, Robotics and Vision*, Singapore (2006)
34. C.W. Hsu, C.C. Chang, C.J. Lin, A practical guide to support vector classification. Technical report, Department of Computer Science and Information Engineering, National Taiwan University, Taipei (2003)
35. H.T. Lin, C.J. Lin, A study on sigmoid kernels for SVM and the training of non-PSD kernels by SMO-type methods. Technical report, Department of Computer Science and Information Engineering, National Taiwan University, Taipei (2003), http://www.csie.ntu.edu.tw/cjlin/papers/tanh.pdf
36. J.C. Platt, Sequential minimal optimization: a fast algorithm for training support vector machines. Technical report, Microsoft Research (1998)
37. J.R. Quinlan, Induction of decision trees. Mach. Learn. **1**(1), 81–106 (1999)
38. S.R. Safavian, D. Landgrebe, A survey of decision tree classifier methodology. IEEE Trans. Syst. Man Cybern. **21**(3), 660–674 (1991)
39. A. Ittner, M. Schlosser, Non-linear decision trees – NDT, in *Proceedings of the 13th International Conference in Machine Learning*, Bari (1996)
40. E.T. Jaynes, Information theory and statistical mechanics. Phys. Rev. **106**(4), 620 (1957)
41. J.M. Bernardo, A.F.M. Smith, *Bayesian Theory* (Wiley, Palo Alto, 1994)
42. K. Trivedi, *Probability and Statistics with Reliability, Queuing and Computer Science Application*, 2nd edn. (Wiley, New York, 2001)
43. J.O. Berger, *Statistical Decision Theory and Bayesian Analysis*, 2nd edn. (Springer, Berlin/Heidelberg, 1985)
44. C. Suen, L. Lam, Multiple classifier combination methodologies for different output levels, in *Multiple Classifier Systems, First International Workshop*. Volume 1857 of Lecture Notes in Computer Science (Springer, Berlin/Heidelberg, 2000)
45. J.O. Berger, *A Mathematical Theory of Evidence*, 2nd edn. (Princeton University Press, Princeton, 1976)
46. J.R. Boston, A signal detection system based on Dempster-Shafer theory and comparison to fuzzy detection. IEEE Trans. Syst. Man Cybern. Appl. Rev. **30**(1), 45–51 (2000)
47. H. Wu, M. Siegel, R. Stiefelhagen, J. Yang, Sensor fusion using dempster-shafer theory, in *IEEE Instrumentation and Measurement Technology Conference*, Anchorage (2002)
48. J.L. Jesneck, L.W. Nolte, J.A. Baker, C.E. Floyd, J.Y. Lo, Optimized approach to decision fusion of heterogeneous data for breast cancer diagnosis. Med. Phys. **33**(8), 2945–2954 (2006)
49. E.I. Gokce, A.K. Shrivastava, J.J. Cho, Y. Ding, Decision fusion from heterogeneous sensors in surveillance sensor systems. IEEE Trans. Autom. Sci. Eng. **8**(1), 228–233 (2006)
50. D. Basak, S. Pal, D. Chandra, Support vector regression. Neural Inf. Process. Lett. Rev. **11**(10), 203–224 (2007)
51. S.K. Aggarwal, L.M. Saini, A. Kumar, Electricity price forecasting in deregulated markets: a review and evaluation. Int. J. Electr. Power Energy Syst. **31**(1), 13–22 (2009)
52. B. Dong, C. Cao, S.E. Lee, Applying support vector machines to predict building energy consumption in tropical region. Energy Build. **37**(5), 545–553 (2005)
53. A. Smola, B. Schölkopf, Statistics and computing. Tutor. Support Vector Regres. **14**(3), 199–222 (2004)

Chapter 6
Analysis and Prediction of Enterprise Service-Level Performance

In addition to production workflow optimization (Chap. 3), the analysis and prediction of the execution performance of an individual process (Chap. 4), and the analysis and prediction of the fulfillment performance of an individual order (Chap. 5), the EIS should further be able to analyze and predict enterprise service-level performance. The collaborative goal of the preceding operation-optimization and knowledge-discovery efforts is to assist an enterprise to achieve superior service-level performance. The eventual concern of an enterprise is to satisfy service-level agreements (SLAs), to provide quality of service guarantees, and to ensure profitability. Therefore, in this chapter we study the analysis and prediction of enterprise service-level performance.

An enterprise service-level performance time series is a sequence of data points that quantify demand, throughput, average order-delivery time, quality of service, or end-to-end cost. Analytical and predictive models of such time series within an EIS can provide meaningful insights into potential business problems and generate guidance for appropriate solutions. Time-series analysis includes periodicity detection, decomposition, and correlation analysis. Time-series prediction can be modeled as a regression problem to forecast a sequence of future time-series data points based on the given time series. The state-of-the-art (baseline) methods employed in time-series prediction generally apply advanced machine-learning algorithms. In this chapter, we propose a new univariate method in dealing with mid-term time-series prediction. The proposed method first analyzes the hierarchical periodic structure in one time series and decomposes it into trend, season, and noise components. By discarding the noise component, the proposed method only focuses on predicting repetitive season and smoothed trend components. As a result, this method significantly improves upon the performance of baseline methods in mid-term time-series prediction. Moreover, we propose a new multivariate method in dealing with short-term time-series prediction. The proposed method utilizes cross-correlation information derived from multiple time series. The amount of data taken from each time series for training the regression model is determined by results

© Springer International Publishing Switzerland 2015

Q. Duan et al., *Data-Driven Optimization and Knowledge Discovery for an Enterprise Information System*, DOI 10.1007/978-3-319-18738-9_6

from hierarchical cross-correlation analysis. Such a data-filtering strategy leads to improved algorithm efficiency and prediction accuracy. In conclusion, by combining statistical methods with advanced machine-learning algorithms, we have achieved a significantly superior performance in both short-term and mid-term time-series predictions compared to state-of-the-art (baseline) methods.

The rest of this chapter is organized as follows. Section 6.1 discusses the background and motivation for enterprise service-level performance analysis and prediction. Section 6.1.4 introduces the state-of-the-art solutions and our data source. In Sect. 6.2, we present hierarchical time-series decomposition, the baseline model, and the proposed model for making mid-term time-series prediction. Following it, Sect. 6.3 presents time-series correlation analysis and the proposed multivariate short-term time-series prediction model used for short-term prediction. Section 6.4 concludes the chapter.

An enterprise information system (EIS) is an intelligent real-time embedded system within an enterprise. For system-level design, optimization, and evaluation, performance prediction at different levels is crucial in almost any operational environment [1]. Such predictions are very informative for making decisions or planning schedules. For example, response performance prediction is required in communication system design and optimization [2], thermal performance prediction is important for scalable design of modern processors [3], temperature and power performance prediction is needed for 3D-chip multiprocessors [4], and test-cost prediction is necessary for integrated and heterogeneous systems [5].

Enterprise service-level objectives (SLOs) include demand, throughput, order delivery time, quality of service, and end-to-end cost [6]. They are usually dominated by on-time delivery or due-date constraints [7]. An enterprise service-level performance time series is a sequence of data points on one of its SLOs [8]. There are many key purposes for analyzing and predicting enterprise service-level performance: (1) to identify and adapt to changing demand patterns; (2) to support the generation of synthetic demand that represents future demand, e.g., demand for several days, weeks or months into the future to support capacity planning exercises; (3) to derive models that can be used for forecasting purposes; and (4) to study how demand influences other service-level performances such as throughput and on-time/late delivery [9]. Analytical and predictive models of such time series can provide meaningful insights into potential business problems and generate guidance for appropriate solutions. Traditionally, it is sufficient to make performance predictions by relying on pre-designed rule-based specification or even human judgment. However, increased system complexity and heterogeneity drives the need for advanced algorithms and real-time performance prediction in embedded systems [10–12].

Time-series analysis includes periodicity detection, decomposition, and correlation analysis. Periodicity of a time series can be identified through classical Fourier transform-based spectral analysis [13] and auto-correlation analysis [14]. Studies in [15] have successfully demonstrated the integration of these two approaches for periodicity detection. Even more sophisticated algorithms are introduced in [16] for the same purpose. Periodicity can be used for time-series decomposition.

A season-trend decomposition procedure based on the LOESS moving-average algorithm was introduced in [17]. The procedure decomposes a time series into additive trend, season, and remainder. Multiple interdependent time series are becoming more correlated with each other; this is the main motivation for performing multivariate analysis for multiple time series.

Time-series predictions mainly involve short-term and mid-term predictions. Short-term prediction focuses on predicting only a few future data points for a given time series. In contrast, mid-term prediction forecasts a sequence of future data points, including at least one period of the given time series. Statistical methods have often been adopted in time-series prediction. Historically, enterprise time-series prediction has relied upon curve fitting and queueing models. Unfortunately, these approaches are typically human-intensive and expensive processes [9, 18]. They are difficult to automate and do not scale well for broader use. In addition, enterprise time-series data are inherently noisy and nonstationary [6]. The nonstationary characteristic implies that the distribution of the time series changes over time. It becomes impossible to statically model the gradual changes in a time series. Techniques for short-term predictions often use approaches such as ARMA [14] or GARCH [19] models. While these approaches may be adequate for the short-term, their predictions quickly converge undesirably to a mean value for longer terms [9]. A review in [20] discusses more methods and progress made in time-series forecasting.

In recent years, machine-learning algorithms have been developed to address time-series or sequential data-prediction problems. Popular algorithms include sliding window methods, hidden Markov models, conditional random fields, and graph transformer networks, and so on [21]. Neural networks have been successfully used for modeling financial time series [22]. Neural networks are universal function approximators that can map any nonlinear function without a priori assumptions about the properties of the data [23]. Unlike traditional statistical models, neural networks are data-driven, nonparametric models. They are also more noise-tolerant in terms of having the ability to learn complex systems with incomplete and corrupt data. Among different machine-learning algorithms, support vector machines (SVMs) have gained traction in time-series prediction applications ranging from the original application in pattern recognition [24, 25] to the extended application of regression estimation [26]. SVMs usually achieve higher generalization performance than traditional neural networks. Training an SVM is equivalent to solving a linearly constrained quadratic programming problem so that the solution is always unique and globally optimal. The remarkable characteristics of SVMs such as good generalization performance, the absence of local minima, and sparse representation of the solution, have contributed their popularity. In summary, state-of-the-art time-series prediction methods use either statistical methods or machine-learning algorithms.

However, univariate mid-term prediction and multivariate short-term prediction are still bottlenecks for real-time responsiveness in an enterprise system [27]. Short-term and mid-term predictions of enterprise service-level performance are difficult to make due to the following reasons. First, there exists a large amount

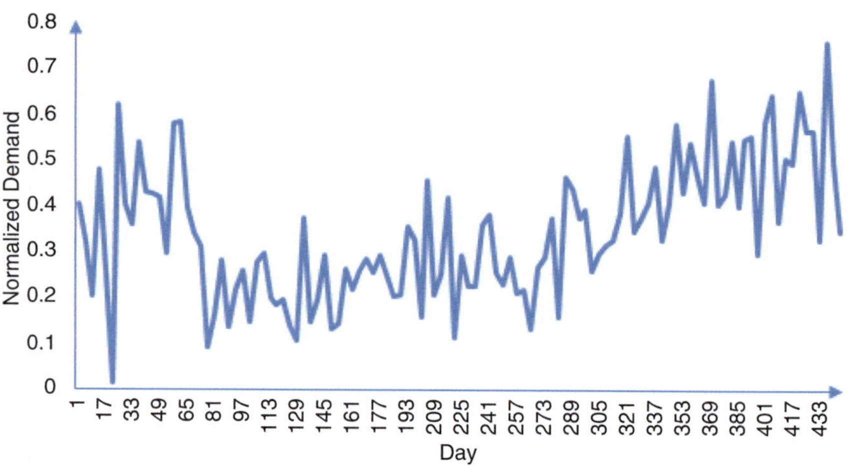

Fig. 6.1 Normalized demand of print service provider X

of variations in multiple domains, e.g., diverse service requests that arrive in a bursty or unpredictable manner, processes that rely on a significant amount of manual work, and the occurrence of stochastic failures. Client demands can also vary dramatically, e.g., "peak-to-peak" ratios may differ by an order of magnitude or more [28]. Moreover, due to the lack of adequate system understanding, some key features might be missed during forecasting. Finally, computationally expensive offline optimization solutions cannot be deployed in real-time applications [27].

Take the digital-print industry as an example. Today, print service providers (PSPs) rely on skilled managers for service-level time-series prediction, which is an error-prone process. It impedes a thorough understanding of the enterprise and the subsequent capacity-planning process. Consider a scenario in which the demand experienced by a PSP X is as shown in Fig. 6.1. If a manager of X was given the task of forecasting the future demand based on this data, it would be difficult for him to understand the complex patterns in it and this could lead to inaccurate predictions. Therefore, there is a need for techniques that can accurately analyze and predict enterprise service-level performance.

In this chapter, we propose a univariate mid-term time-series prediction method and a multivariate short-term time-series prediction method. The proposed methods account for a hierarchical periodic structure in any given time series and temporal correlations between different time series. By combining statistical methods and advanced machine-learning algorithms, we have achieved improved performance in both short-term and mid-term time-series predictions as opposed to state-of-the-art (baseline) solutions.

The main contributions of this chapter are as follows.

- Hierarchical season-trend decomposition enables a better understanding of the time-series structure. Season components tend to repeat themselves, whereas noise components are highly unpredictable. By identifying and discarding noise components, the proposed univariate mid-term prediction method only focuses on predicting repetitive season and smoothed trend components. This method has significantly improved upon the performance of state-of-the-art (baseline) methods in mid-term time-series prediction.

- Studying components obtained from hierarchical decomposition also helps us to perform and validate cross-correlation analysis between multiple time series. Cross-correlation information reveals a significant amount of cause-and-effect relationships between different time series.
- The proposed multivariate short-term time-series prediction method uses cross-correlation information explored from multiple time series. The amount of historical data of each time series used for training the regression model is determined through hierarchical cross-correlation analysis. The inclusion of only necessary information improves both algorithm efficiency and prediction accuracy.

6.1 Problem Statement, Baseline Methods, and Data Source

6.1.1 Problem Statement

We can separate a time series into two parts: known data (including historical and current) and predictive data. As illustrated in Fig. 6.2, suppose that given a known time-series data $f(t)$, $f(t-1)$,..., and $f(t-L)$ at time t, $t-1$, ..., and $t-L$, respectively, we develop models that can make an n-step look-ahead prediction (i.e., predict the future values of $f(t+1), f(t+2), \ldots, f(t+n)$) by analyzing known data. The constant L is the length of the known time series (the number of historical data points). The parameter n is the furthest data point in the future that is of interest for prediction.

Relative squared error (RSE) is performance index that is commonly used to evaluate time-series models [29]. It is defined in Eq. (6.1) as follows.

$$\text{RSE} = \frac{\sum_{i=1}^{N} \left(Y_i^* - Y_i \right)^2}{\sum_{i=1}^{N} \left(\overline{Y}_i - Y_i \right)^2} \tag{6.1}$$

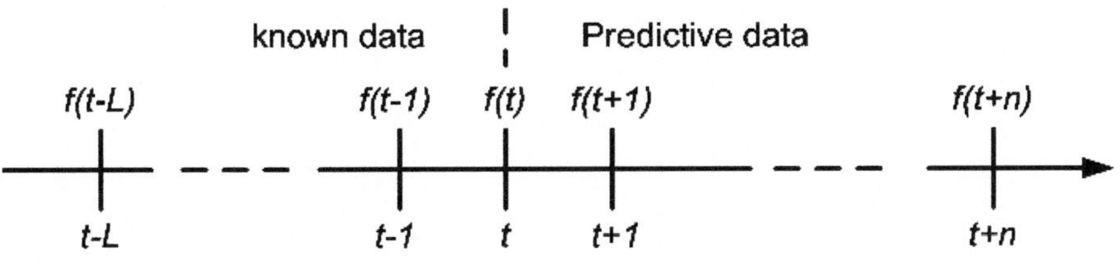

Fig. 6.2 Time-series prediction model

where Y_i is the actual value, $\overline{Y_i}$ is the mean of actual value, Y_i^* is the modeled or predicted value, and N is the total number of data points that have been evaluated. RSE can be compared between time series whose values are measured in different units.

There are two methods to make an n-step ahead prediction. Note that only one future data point is predicted each time. Once a data point is predicted, it will be treated as a known data point and used as input to predict the next data point. One method is iterative prediction, in which whenever a data point is predicted, the model will be re-trained by adding the predicted data as known data. For an n-step ahead iterative prediction, n models need to be trained to make all the predictions. The other method is one-time prediction, i.e., all the future data points are predicted by the model which was initially trained by the original time series. In general, iterative prediction makes less error than one-time prediction. However, the quality of the results deteriorates very quickly as the predictive horizon increases due to the accumulation of prediction errors, regardless of the prediction algorithm [30].

6.1.2   Baseline Univariate Method

A data-driven method for single time-series prediction can be characterized as a univariate method using existing data points as input to predict future data points for the time series. For a machine-learning based univariate time-series prediction method, each input instance for training the model is a vector (x_i, y_i), where $x_i \in X \subseteq R^m$, m is the number of features, and y is the actual value. The training data includes multiples of such instances. The generation of the initial training data for a time series with L known data points is illustrated in Fig. 6.3. In order to predict one data point, a sequence of g historical data points should be provided. In this case, $m = g$. Parameter g is referred to as the lag of the model. It can be seen that the number of instances in the training data is $L - g$. Theoretically, the longer the known time series (L) and the more the historical data (g) in each instance, the more informative is the training data.

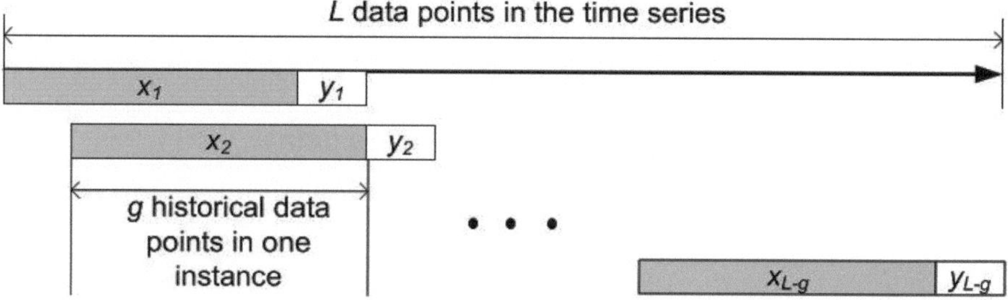

Fig. 6.3 Generation of the training data for an univariate time-series prediction method

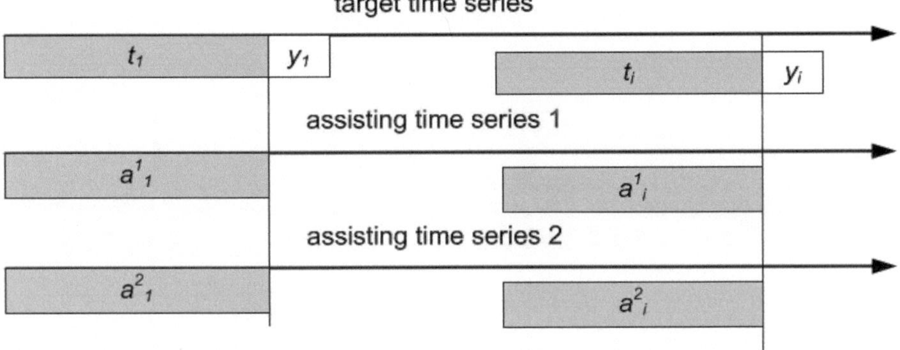

Fig. 6.4 Generation of the training data for a multivariate time-series prediction method

6.1.3   Baseline Multivariate Method

Compared with the univariate method illustrated in Fig. 6.3, the multivariate method involves one target time series and multiple assisting time series as shown in Fig. 6.4. This method uses the information of the target and multiple assisting time series in order to predict new data for the target time series. Note that the input feature vector x_i now includes g-lag data from the target and the assisting time series, i.e., $x_i = (t_i, a_i^1, a_i^2, \ldots)$, $1 \leq i \leq (L - g)$, where L is the length of the target time series, parameter g is the lag of the model, t is the historical data of the target time series, and a^j is the historical data of the jth assisting time series.

6.1.4   Data Source

The IT stack (enterprise resource planning, manufacturing execution and workflow management systems) in an EIS tracks the lifecycle of an order and documents the production activities (e.g., receiving, negotiating, admitting, planning, realizing, and delivering an order). Each tracking event produces an event log written to an event-log database. An event log includes a timestamp, a process ID, an equipment ID, an employee ID, IDs for this order, this product and this part, and other ancillary information. The factory managers use the production event logs to debug the production process in case things go wrong. Our experimental environment is composed of an event-log database obtained from a large multi-national mass customization enterprise. It includes 664 million records. It is managed by a 5-node VERTICA cluster (v6.1.1-0, Ubuntu 64 bit) in support of fast computation. We mine this production event-log database to uncover the production reality and extrapolate the trend into the future. In compliance with our confidentiality agreement with the enterprise, real-life data have been normalized by nonlinear functions (two sine functions) for the purpose of visualization. Therefore, while no actual value of time or quantity is presented, the experimental results are for realistic scenarios.

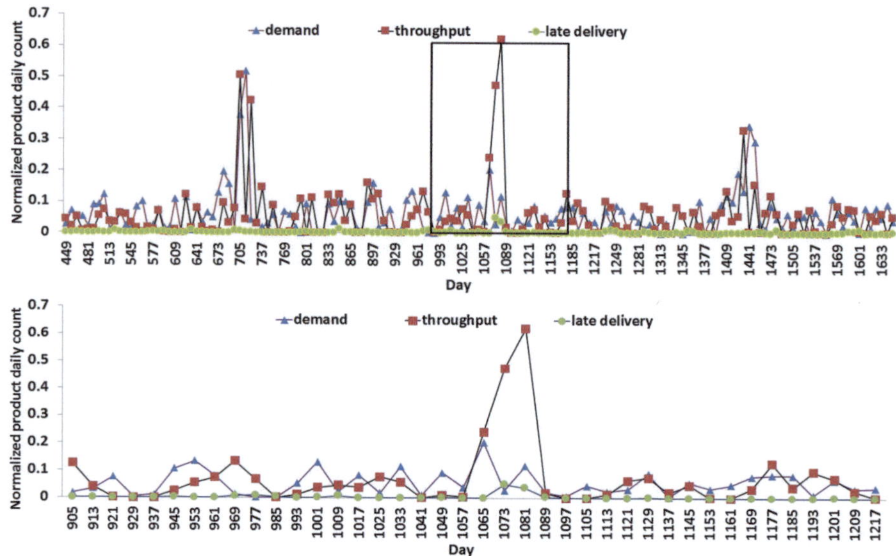

Fig. 6.5 Data extracted from a commercial enterprise's production event-log database. The *first row* shows the daily counts of demand, throughput, and late delivery. The *second row* zooms into the region covered in the *black box* of the first row

We have extracted data from a commercial enterprise's production event-log database from the year 2008 to 2013. Figure 6.5 shows a part of the extracted data. The first row shows the flows of the demand, production throughput, and late delivery in terms of daily product count. The second row zooms into the region inside the black box on the first row. It reveals the weekly periodic patterns that cannot be determined from the lower-resolution view in the first row.

In addition, a commercial enterprise's demand can be further classified into four product types. The first and second rows in Fig. 6.6 show different product type percentage counts among daily counts of demand and late delivery, respectively. We partition the data into training (time series before March 2012) and testing data (time series between April 2012 to April 2013) to train and test the proposed methods.

6.2   Mid-Term Time-Series Analysis and Prediction

Mid-term time-series analysis and prediction are intended for system-level design-time analysis, for example, for capacity planning, personnel management, and business policy decisions.

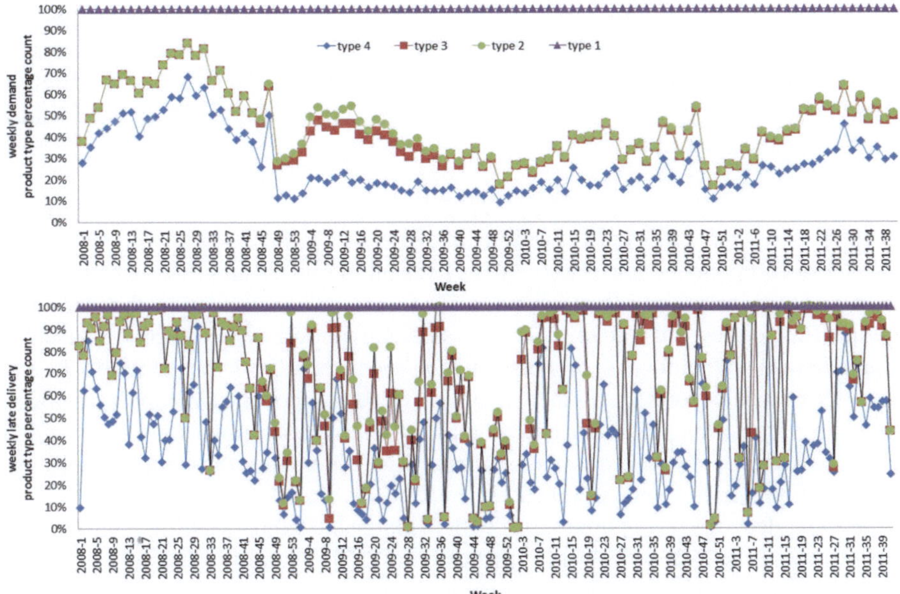

Fig. 6.6 Data extracted from a commercial enterprise's production event-log database show different product type percentage counts. The horizontal axis is year-week

6.2.1 Time-Series Decomposition and Modeling

Enterprise time-series data often contain multiple periodic patterns. Daily, weekly, monthly, quarterly and yearly patterns can be observed at different scales. For instance, daily patterns distinguish peak and off-peak hour and weekly patterns distinguish weekday and weekend trend, while yearly patterns distinguish holidays and work days. It is obvious that understanding this time-varying feature is the key to enterprise time-series modeling.

Classical Fourier transform-based spectral analysis can be used to identify time-series periodicity [13]. A Fourier transform gives an overlay of harmonics for the time-varying magnitude of the time series. The most dominant frequencies (frequencies showing large intensities) provide candidate periods. Auto-correlation analysis can also be used to identify periodicity [14]. Auto-correlation reveals similarity between a time series and its shifted version by different lags. A high value for the auto-correlation at lag g denotes that the original time series looks similar to itself when shifted by g time steps. Therefore, if the auto-correlation shows local extrema at multiples of a lag g, it is a strong indicator that g is a candidate period. Cross-correlation can be used to identify similarity among different time series. Formal definition and further details on auto-correlation are presented in [14]. Since there are many standard techniques for periodicity detection, we directly use candidate periods obtained from observing the data and confirmed by these techniques.

We decompose the time series with respect to each identified candidate period. According to a classical time-series additive model [17], a time series consists of a trend component, a season component, and a remainder. Suppose the time-series data, the trend component, the season component, and the remainder are denoted by Y_v, T_v, S_v, and R_v, respectively, for $v = 1$ to N. Then

$$Y_v = T_v + S_v + R_v \qquad (6.2)$$

The trend component is the low-frequency variation in the time-series data together with nonstationary long-term changes. The season component is the variation in the time-series data at or near the seasonal period. The remainder usually characterizes the influence of noise, which corresponds to the variation in the time-series data beyond that in the season and trend components. We apply a season-trend decomposition procedure based on LOESS moving-average algorithms (STL) as discussed in [17] on a commercial enterprise's demand, throughput, and late-delivery time series shown in Fig. 6.5.

We first decompose the time series with the largest candidate period (i.e., a yearly period) using the STL procedure as presented in [17]. The resulting inter-year trend components are shown in Fig. 6.7. Seasonal components are repeated over yearly period. Figure 6.8 shows the intra-year season components, in which weekly pattern and burstiness are revealed more clearly. The remainders have mean values very close to zero, therefore, the remainders were treated as white noise and discarded.

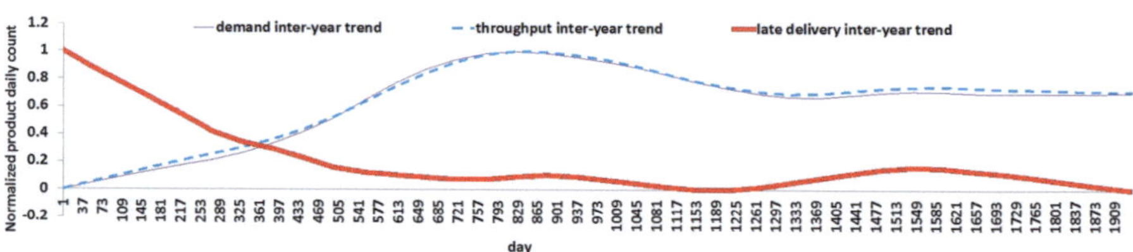

Fig. 6.7 Time series in Fig. 6.5 are decomposed to inter-year trend components

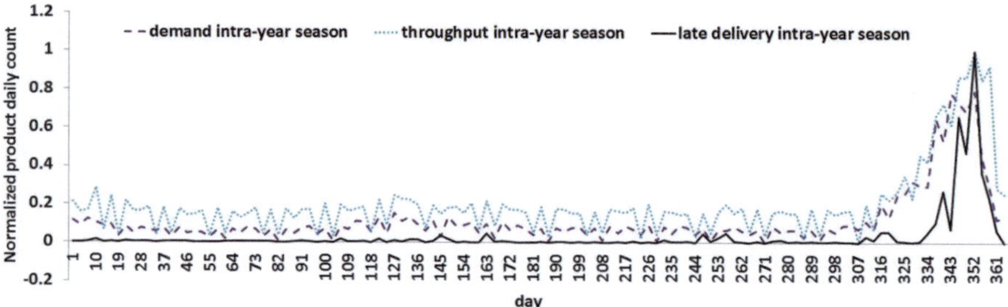

Fig. 6.8 Time series in Fig. 6.5 are decomposed to intra-year season components

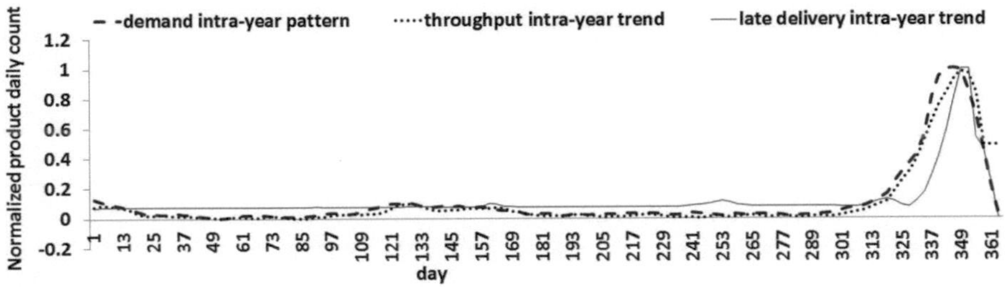

Fig. 6.9 Intra-season components in Fig. 6.8 are decomposed to intra-year trend components

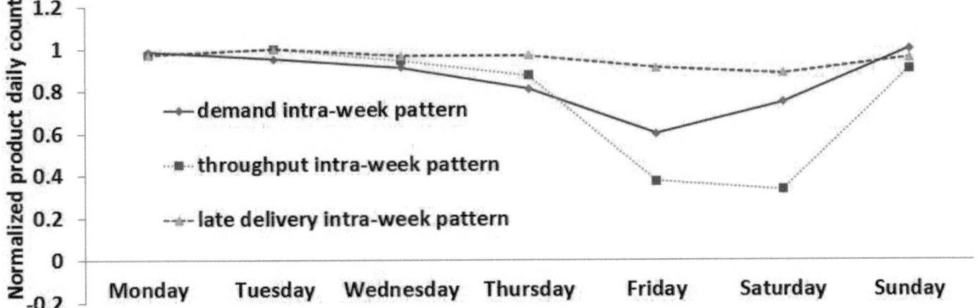

Fig. 6.10 Intra-season components in Fig. 6.8 are decomposed to intra-week season components

Since weekly periodicity is the next smaller candidate period with strong intensity in spectral analysis, we further apply the STL procedure to decompose the intra-year season component (shown in Fig. 6.8) to intra-year trend component (shown in Fig. 6.9) and intra-week season component (shown in Fig. 6.10).

Although demonstrated through a commercial enterprise's data, the same hierarchical decomposition procedure can be applied to any time-series data. In summary, hierarchical time-series decomposition involves the following steps: (1) hierarchical periodicity detection through human inspection, spectral analysis, and auto-correlation methods; (2) first-level STL decomposition starting from the largest candidate period; (3) remaining STL decompositions always start with the next smaller candidate period with strong intensity in spectral analysis. Note that a multiple of a candidate period is also another valid candidate period.

We then model the original time series shown in Fig. 6.5 according to the hierarchical aggregation strategy shown in Fig. 6.11. We aggregate three components into the modeled time series (the first row in Fig. 6.11). The first component (the second row in Fig. 6.11) is the inter-year trend (Fig. 6.7). The second component (the third row in Fig. 6.11) is the yearly-repeated intra-year trend (Fig. 6.9). The third component (the last row in Fig. 6.11) is the weekly-repeated intra-week season (Fig. 6.10). Note that the aggregation of the second and the third components replaces the noisy yearly season obtained from the first-level decomposition procedure. Therefore, the hierarchical aggregation strategy excludes noise obtained from

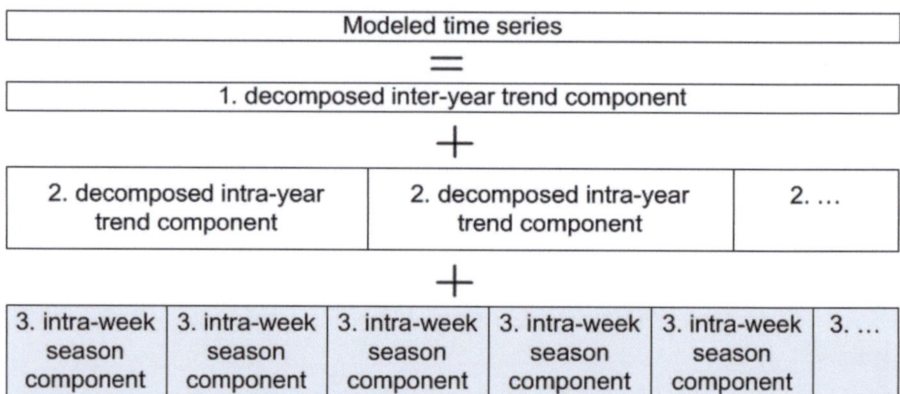

Fig. 6.11 Hierarchical time-series aggregation strategy

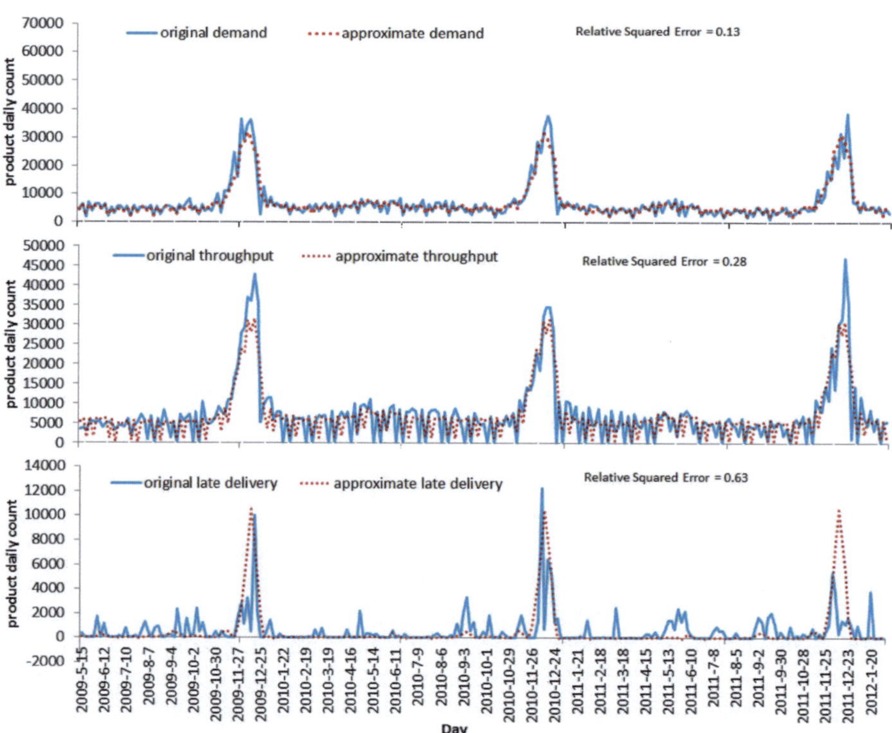

Fig. 6.12 The modeled demand, throughput, and late delivery daily count obtained from the hierarchical decomposition and integration procedures

both the first-level and the second-level decomposition procedures. The modeled time series are shown in Fig. 6.12. Relative squared errors of the modeled demand, throughput, and late delivery time series are 0.13, 0.28, and 0.63, respectively. It can be seen that the demand and throughput time series are better structured than the late

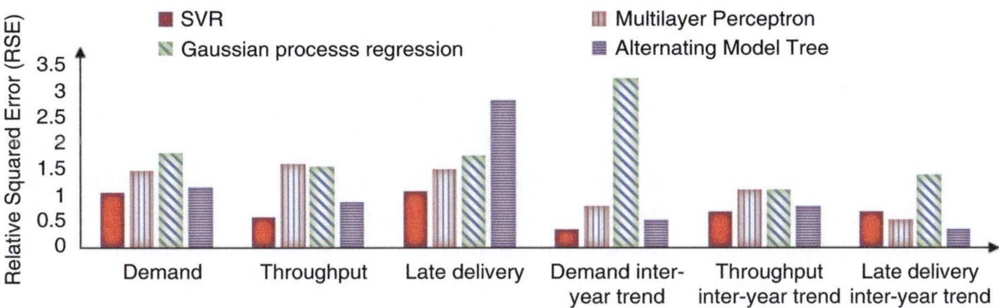

Fig. 6.13 Relative Squared Error (RSE) for different time series data

delivery time series. Consequently, the more structured the original time series, the better the modeled time series approaches the original one.

6.2.2   Support Vector Regression

The application of SVM to regression is called support vector regression (SVR). The tutorial in [31] describes the process of selecting algorithms for training SVR models. We are motivated by the many successful applications of SVR models in the design automation literature [32, 33] (also discussed in Sect. 6). In order to justify our choice, we also compared the prediction accuracy of SVR to three other regression algorithms:

- *Multi-layer Perceptron* is the most popular type of artificial neural networks (ANNs) because of its ability to model non-linear and noisy data. In this work, we have used a multi-layer perceptron with two hidden layers.
- *Gaussian Process Regression* is a probabilistic model which assumes that the training data can be represented as a sample from a multivariate Gaussian distribution.
- *Alternating Model Tree* proposed by [34] is a alternating decision tree based model for regression problems.

Figure 6.13 shows the comparison results for different time series data. It can be seen that SVR achieves the highest prediction accuracy for our testing data as compared to other regression algorithms in all but one case.

There are many parameters that must be set with SVR. In our experiments, we have experimented with several combinations and finally adopted a Radial Basis Function (RBF) with $\gamma = 0.1$ as the kernal, tube size $\varepsilon = 0.01$, and the regularization factor $C = 100$ [31].

6.2.3   Implementation of Baseline Methods

The baseline univariate method is introduced in Sect. 6.1.2. Today's data-driven time-series prediction approaches (e.g., WEKA [35]) treat the data as being flat and rely on machine-learning algorithms (e.g., SVM) to uncover the patterns. The class of approaches that rely on extensive manual tuning of model parameters can achieve high accuracy in short-term predictions. However, as the predictive horizon increases, performance deteriorates. This happens because overfitting of the known time series is beneficial for making short-term predictions but detrimental for making longer-term predictions [30].

Theoretically, as discussed in Sect. 6.1.2, the longer the known time series (L) and the more the historical data in each instance (g), the more informative is the training data. However, the feasible values of L and g are limited in real-time applications because the training time of an SVR scales as a function that is between quadratic and cubic with respect to $L - g$ (the number of training samples) and g (the number of input features) [36].

The time required for training an SVR model is the main contributor to delay in a real-time application. Since demand time series is the most structured data, we use demand data to evaluate time required for training an SVR model. Experimental results on the average time to train the model and to make predictions are shown in Tables 6.1 and 6.2. It can be clearly seen that new predictions can be made instantly when the one-time prediction model is used. On the other hand, the time required to make new predictions using the iterative model increases linearly with the predictive horizon. Therefore, it is not practical to use the iterative model when the predictive horizon is large.

Figure 6.14 shows the comparison results for different baseline methods. It can been seen that baseline methods with less lags cannot capture the yearly pattern of the data. On the other hand, iterative baseline methods with a 1-year lag are infeasible for real-time applications. Therefore, we select the 1-year lag one-time model as the baseline method to compare the performance of our proposed method, as discussed in Sect. 6.2.4.

Table 6.1 Time required for building the iterative prediction model with a Windows workstation (8G memory and 2.5 GHz CPU)

Time required (s)	Iterative						
	1 step ahead	20 step ahead	40 step ahead	80 step ahead	160 step ahead	320 step ahead	420 step ahead
Lag = 30	11	239	481	992	2,076	4,397	6,146
Lag = 60	12	266	542	1,300	2,551	4,780	5,130
Lag = 120	10	272	482	939	1,873	3,889	5,228
Lag = 240	8	178	364	879	1,521	4,096	6,033
Lag = 364	12	190	431	843	1,689	5,026	6,188

Table 6.2 Time required for building the one-time prediction model with a Windows workstation (8G memory and 2.5 GHz CPU)

Time required (s)	One-time						
	1 step ahead	20 step ahead	40 step ahead	80 step ahead	160 step ahead	320 step ahead	420 step ahead
Lag = 30	11	11	11	11	11	11	11
Lag = 60	12	12	12	12	12	12	12
Lag = 120	10	10	10	10	10	10	10
Lag = 240	8	8	8	8	8	8	8
Lag = 364	12	12	12	12	12	12	12

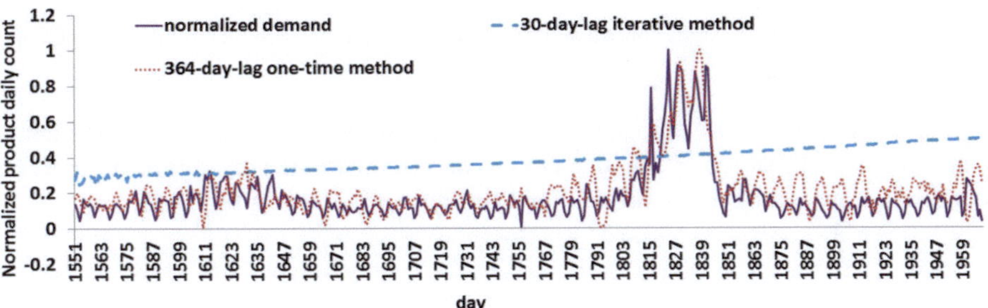

Fig. 6.14 Predicted demand by different baseline methods

6.2.4 Proposed Univariate Mid-Term Time-Series Prediction Method

As discussed in Sect. 6.2.3, the main difficulty in mid-term time-series prediction is the deteriorating performance as the predictive horizon increases. The main motivation for the hierarchical decomposition of a time series is to explore its structure more thoroughly and reduce the unpredictable components. Instead of predicting the whole data, which contains white noise, we investigate distributed predictions of each trend component and then reconstruct the whole data by integrating predicted trend components and repeated season components.

We used SVR models to learn and predict the data for the next year. According to the hierarchical decomposition procedure shown in Sect. 6.2.1 and the aggregation procedure shown in Fig. 6.11, we predict the data for the next year by aggregating the predicted data of three components as illustrated in Fig. 6.15. The first component is the predicted inter-year trend for the next year. The second component is the historical intra-year trend. The historical intra-week season component in Fig. 6.10 is repeated for 52 times as the third component. Therefore, the predicted data for the next year is the aggregation of the predicted inter-year trend, the historical intra-year trend, and the repeated intra-week season. This is because after the two-level decomposition, the intra-year season component is replaced by the

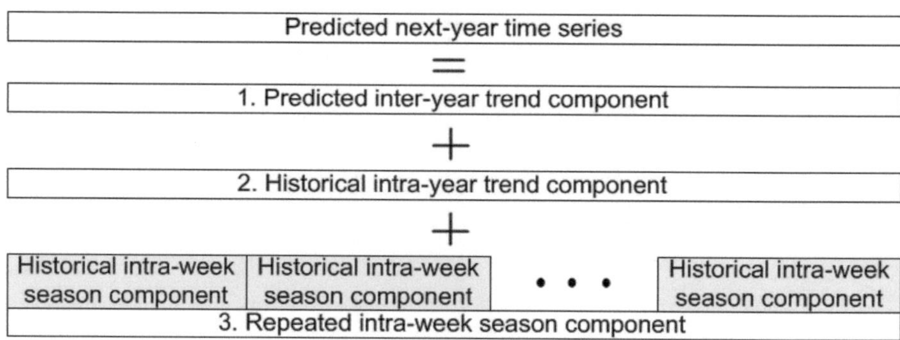

Fig. 6.15 Predicted time series constructed from predicted and repeated hierarchical components

aggregation of the intra-year trend and the repeated intra-week season components. The complete time series for the next year is constructed by aggregating each component according to the hierarchical structure as shown in Fig. 6.15.

6.2.5 Results and Discussions

Figure 6.16 shows the comparison between the proposed method and the baseline method. Relative square errors (RSEs) of 1-year-ahead predicted demand and throughput given by the proposed method are 0.23 and 0.32, respectively. The corresponding errors are 0.41 and 0.94 for the baseline method. Evaluated using RSE, the proposed method demonstrates better performance than the baseline method. Furthermore, it can be seen that for the throughput time series, which is less structured than the demand data, the baseline method deteriorates significantly as the predictive horizon increases. Therefore, we can predict scenarios for the least structured late-delivery data. The baseline method clearly cannot be used for predicting 1-year-ahead throughput and late delivery due to large errors. The improvement for such less structured data through the proposed method is more significant. The proposed method demonstrated its robustness for less structured time-series data, which is a major bottleneck for the baseline method.

Other than providing a thorough understanding of the periodic nature of a time series, the improved performance gained by the proposed method also demonstrates another motivation for hierarchical time-series decomposition. The amount of noise that is removed in the decomposition is proportional to the number of hierarchy levels involved in the process. This is due to the fact that noise is the random variation in a time series and it cannot be described in terms of trend or seasonality. However, the number of levels of decomposition should not be arbitrarily increased. It should be applied as long as it can better abstract the trend and seasonality.

Furthermore, another advantage of hierarchical decomposition is that it helps us to improve n-step ahead prediction when n is even larger than the largest period

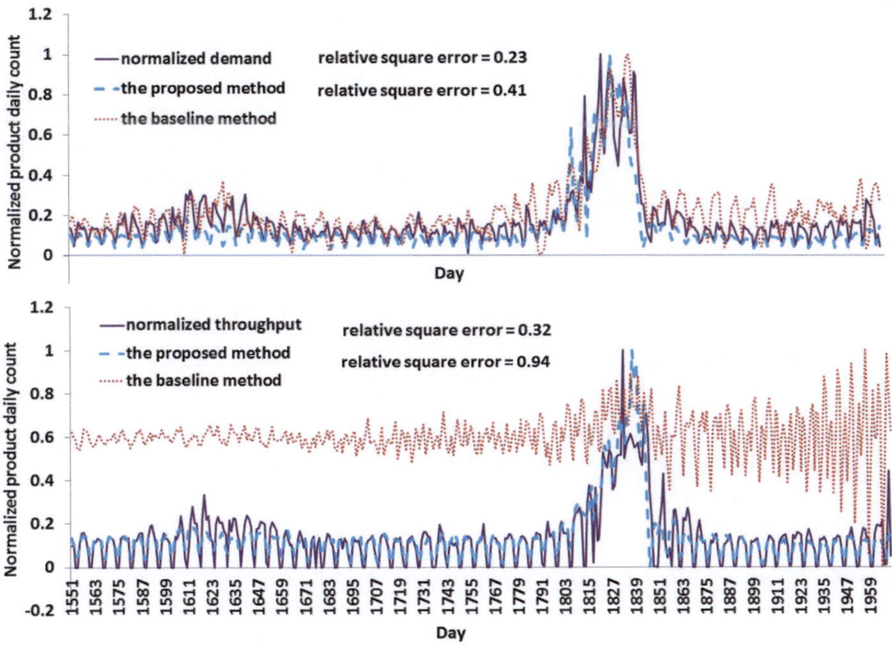

Fig. 6.16 Predicted demand and throughput by different models

Table 6.3 Relative squared error for various publicly available time series data sets

Relative Squared Error	Monthly milk production Jan 62 – Dec 75	Monthly chocolate production Jul 57 – Aug 95	Quarterly farm production Sep 59 – Mar 93	Monthly river flow Jan 31 – Dec 78
Baseline method	0.49	2.24	0.09	0.26
Proposed method	0.23	0.65	0.05	0.17

P_{max}. In such a case, the original time series should be initially decomposed by a multiple factor of its largest period such that $c \cdot P_{max} \geq n$, where c is a constant. Meanwhile, the largest period P_{max} is updated to a multiplication of the original value, $P'_{max} = c \cdot P_{max}$. It is an advantageous feature for time-series data with smaller periods but require longer-term predictions. Considering the commercial enterprise's data, if weekly period was its largest detected period, we can still first decompose it with a yearly period since yearly period is a multiplication of weekly period.

We also applied the proposed technique to publicly available time series data sets [37] to demonstrate its generality. Table 6.3 shows the comparison between the proposed method and the baseline method for different time series data sets. Since each of these time series data sets have a single period, they are decomposed into the trend, season, and noise components only once. To find the predicted time series, we follow a procedure similar to the one shown in Fig. 6.15. In this case, the predicted time series is just the sum of the predicted trend and the historical season component. The baseline method used here is same as the one described in Sect. 6.2.3. It can be seen from Table 6.3 that the proposed method performs better than the baseline method for all the data sets. Therefore, the decomposition-based prediction method generally performs better than the tradional method.

6.3 Multivariate Short-Term Time-Series Analysis and Prediction

Short-term time-series analysis and prediction are intended for run-time analysis, for example, real-time job scheduling and resource allocation.

6.3.1 Time-Series Cross-Correlation Analysis

The decomposition procedure and the hierarchical patterns of the demand, through-put and late-delivery curves led us to examine the cross correlations between them. There are clear lags of peaks in throughput and late delivery compared to demand as shown in Fig. 6.5.

For modeled integrated time series (Fig. 6.12), intra-week season pattern (Fig. 6.10), intra-year trend pattern (Fig. 6.9), and inter-year trend pattern (Fig. 6.7), we compute cross correlations among two time series. Cross correlation identifies the relationship among two time series X_v and Y_v. When $X_{v\pm1}$, $X_{v\pm2}$, ..., and X_{v+g} are predictors of Y_v, the lag g is defined as the cross correlation. When lag g is negative, it means that X_v leads Y_v. When lag g is positive, it means that X_v lags Y_v or Y_v leads X_v. The steps for performing cross-correlation analysis among two time series can be found in [14]. Results in Table 6.4 show that in general demand leads throughput and late delivery by 1–3 days. For example, in the first row, the modeled demand leads the modeled throughput by 1 day and leads the modeled late delivery by 3 days. There is no strong correlation found between the modeled throughput and the modeled late delivery. This agrees with our supply-chain empirical knowledge about the commercial enterprise.

Figure 6.17 shows the normalized data in December 2012 by offsetting through-put and late delivery 3 days ahead to provide the visual contrast about such cross-correlation.

Table 6.4 Cross correlation results measured in terms of lag (day as unit) between two time series

Most correlated lag between two time series	Demand & throughput	Demand & late delivery	Throughput & late delivery
Modeled time series	−1	−3	0
Intra-week pattern	0	−1	0
Intra-year pattern	−3	−3	−1
Inter-year pattern	−3	−3	−1

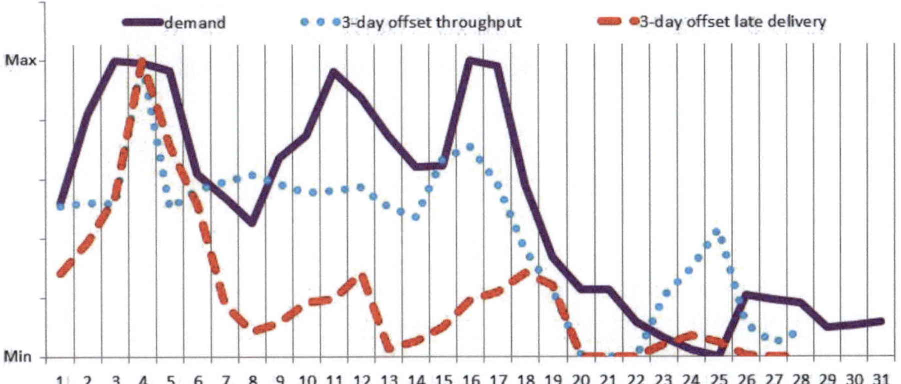

Fig. 6.17 Normalized demand, and 3-day offset throughput and late delivery daily count in December 2012 from the commercial enterprise's data

6.3.2   Implementation of Baseline Methods

The results from cross-correlation analysis motivate us to propose that historical demand, throughput, and late delivery data together can improve the quality of the prediction. To validate this hypothesis, we built a multivariate model that predicts the next data point (one-step ahead) of a time series by utilizing multiple assisting time series.

We used both the standard univariate (Sect. 6.1.2) and multivariate (Sect. 6.1.3) time-series prediction methods implemented in Weka [35] as baseline methods for comparision.

6.3.3   The Proposed Multivariate Short-Term Time-Series Prediction Method

We design a new multivariate short-term time-series prediction method based on the conventional multivariate method (Sect. 6.1.3). Considering that the training time for an SVR model scales between quadratic and cubic with respect to the number of training samples and the number of input features [36], we try to use only the most informative features while guaranteeing the prediction performance.

The generation of the training data for the proposed multivariate method is similar to the conventional method shown in Fig. 6.1.3, with a different lag for the assisting time series data in this case. Rather than including as many historical data points as the target time series, the number of historical data points from the assisting time series is wisely determined. Different weights of assisting time series for making the prediction are reflected through selecting different amounts of historical data points from each assisting time series. The proposed multivariate short-term time-series prediction method uses the cross-correlation analysis results to determine the length of historical assisting time series that should be used for prediction.

6.3.4   *Results and Discussions*

Prediction results after March 2012 for throughput (target time series) using demand as the assisting time series are shown in Table 6.5. It shows that, for all methods, year-long historical throughput gives better results than month-long historical throughput. This is because the largest period of the time series is 1 year. The proposed multivariate method outperforms the baseline methods. In addition, for the proposed multivariate method, using 4/5-day historical demand as the assisting time series gives the highest prediction accuracy. The best results are plotted in Fig. 6.18.

Table 6.5 Relative square errors of different prediction methods for short-term (1-day-ahead) throughput prediction

Method	Regression model parameters	Relative square error	
		Target time series	
		Lag = 364	Lag = 30
Baseline method	Univariant	0.316	0.318
	Multivariate	0.543	0.548
The proposed multivariate method	Assisting time series lag	Target time series	
		Lag = 364	Lag = 30
	6	0.179	0.175
	5	0.174	0.175
	4	0.174	0.175
	3	0.184	0.185
	2	0.179	0.187
	1	0.196	0.199

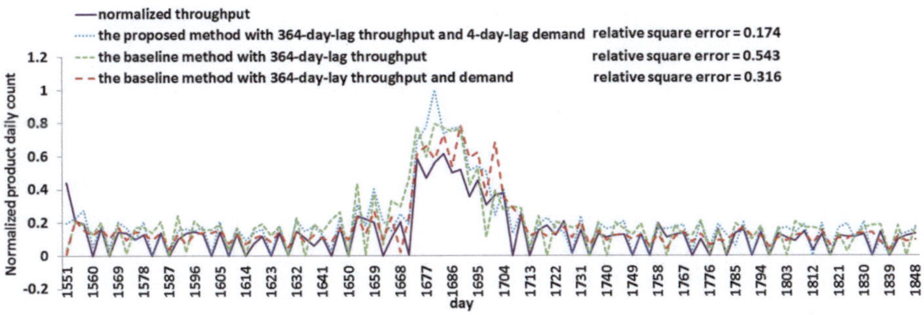

Fig. 6.18 Predicted 1-day-ahead throughput data after 2012 March

We conduct the same experiment for late delivery. The minimum RSE from the baseline methods is 0.713, whereas the best performance of the proposed multivariate method is 0.386. It can be seen that, more data points from the assisting time series cannot guarantee a higher prediction accuracy. Including unnecessary features can even deteriorate model accuracy. Rather than including more data points, lags of the target time series as well as the assisting time series should be selected based on the decomposition and cross-correlation analysis results. Demonstrated through the commercial enterprise's real-life data, the length of the target time series should be its largest period, and the length of an assisting time series should be no less than its number of leading lags with respect to the target time series.

6.4 Conclusion

In this chapter, we have proposed an enterprise service-level performance analysis and prediction system. Prediction results using this system for the commercial enterprise's real-life data have been compared to baseline methods.

It can be seen that, although time-series prediction using machine-learning algorithms have been studied and successfully applied in many applications, the advantages of combining statistical analysis results were not fully investigated. The proposed system has demonstrated the importance of thorough analysis of correlated time series for making mid-term and short-term time-series predictions. It clearly outperforms stand-alone machine-learning based methods.

The advantages of the proposed system and the reasons for superior performance compared to baseline methods can be summarized as follows:

- Hierarchical season-trend decomposition helps us to get a better understanding of the time-series data structure. By performing hierarchical decomposition, the proposed method greatly reduced unpredictable components from the original time-series data by removing noise components. This feature has significantly improved upon the performance of state-of-the-art (baseline) methods in mid-term time-series prediction.

- Hierarchical decomposition results also helps us to perform and validate cross-correlation analysis between different time series. Cross-correlation information reveals cause-and-effect relationships between different time series.

- A multivariate time-series method uses additional useful information, therefore, it outperforms a univariate method. In the proposed multivariate short-term time-series prediction method, cross-correlation results have been used to determine the amount of historical data to be used by the regression model. Providing only the necessary information improves both algorithm efficiency and model accuracy.

References

1. C.L. Dunn, J.O. Cherrington, A.S. Hollander, E.L. Denna, *Enterprise Information Systems: A Pattern-Based Approach*, vol. 3 (McGraw-Hill/Irwin, Boston, 2005)
2. R. Schneider, D. Goswami, S. Chakraborty, U. Bordoloi, P. Eles, Z.B. Peng, Quantifying notions of extensibility in flexray schedule synthesis. ACM Trans. Des. Autom. Electron. Syst. (TODAES) **19**(4), 32:1–32:37 (2014)
3. R. Cochran, S. Reda, Thermal prediction and adaptive control through workload phase detection. ACM Trans. Des. Autom. Electron. Syst. (TODAES) **18**(1), 7:1–7:19 (2013)
4. D.-C. Juan, S. Garg, D. Marculescu, Statistical peak temperature prediction and thermal yield improvement for 3d chip multiprocessors. ACM Trans. Des. Autom. Electron. Syst. (TODAES) **19**(4), 39:1–39:23 (2014)
5. S. Biswas, H.F. Wang, R.D. Blanton, Reducing test cost of integrated, heterogeneous systems using pass-fail test data analysis. ACM Trans. Des. Autom. Electron. Syst. (TODAES) **19**(2), 20:1–20:23 (2014)
6. L.D. Xu, Enterprise systems: state-of-the-art and future trends. IEEE Trans. Ind. Inform. **7**, 630–640 (2011)
7. J. Zeng, I.-J. Lin, E. Hoarau, G. Dispoto, Productivity analysis of print service providers. J. Imaging Sci. Technol. **54**(6), 1–9 (2010)
8. A. Biem, H. Feng, A. Riabov, D. Turaga, Real-time analysis and management of big time-series data. IBM J. Res. Dev. **57**(3/4), 1–8 (2013)
9. D. Gmach, J. Rolia, L. Cherkasova, A. Kemper, Workload analysis and demand prediction of enterprise data center applications, in *IEEE 10th International Symposium on Workload Characterization, 2007, IISWC 2007*, Boston (2007), pp. 171–180
10. R. Rosales, M. Glass, J. Teich, B. Wang, Y. Xu, R. Hasholzner, Maestro–holistic actor-oriented modeling of nonfunctional properties and firmware behavior for mpsocs. ACM Trans. Design Autom. Electron. Syst. (TODAES) **19**(3), 23 (2014)
11. C.-H. Tu, H.-H. Hsu, J.-H. Chen, C.-H. Chen, S.-H. Hung, Performance and power profiling for emulated android systems. ACM Trans. Design Autom. Electron. Syst. (TODAES) **19**(2), 10 (2014)
12. B.H. Meyer, A.S. Hartman, D.E. Thomas, Cost-effective lifetime and yield optimization for NoC-based MPSoCs. ACM Trans. Design Autom. Electron. Syst. (TODAES) **19**(2), 12 (2014)
13. J.W. Cooley, J.W. Tukey, An algorithm for the machine computation of complex Fourier series. Math. Comput. **19**, 297–301 (1965)

14. G.E.P. Box, G. Jenkins, G. Reinsel, *Time Series Analysis: Forecasting and Control*, 3rd edn. (Prentice Hall, Upper Saddle River, 1994)
15. M. Elfeky, W. Aref, A. Elmagarmid, Periodicity detection in time series databases. IEEE Trans. Knowl. Data Eng. **17**(7), 875–887 (2005)
16. Z. Li, J. Han, Mining periodicity from dynamic and incomplete spatiotemporal data, in *Data Mining and Knowledge Discovery for Big Data* (Springer, Heidelberg, 2014), pp. 41–81
17. R.B. Cleveland, W.S. Cleveland, J.E. McRae, I. Terpenning, STL: a seasonal-trend decomposition procedure based on loess. J. Off. Stat. **61**, 3–73 (1990)
18. R.S. Pindyck, D.L. Rubinfeld, *Econometric Models and Economic Forecasts*, vol. 2 (McGraw-Hill, New York, 1981)
19. T. Bollerslev, Generalized autoregressive conditional heteroskedasticity. J. Econ. **31**(3), 307–327 (1986)
20. J.G.D. Gooijer, R.J. Hyndman, 25 years of time series forecasting. Int. J. Forecast. **22**(3), 443–473 (2006)
21. T. Dietterich, Machine learning for sequential data: a review, in *Structural, Syntactic, and Statistical Pattern Recognition*, ser. Lecture Notes in Computer Science, vol. 2396 (Springer, Berlin/Heidelberg, 2002), pp. 15–30
22. J. Moody, J. Utans, Architecture selection strategies for neural networks: application to corporate bond rating prediction, in *Neural Networks in the Capital Markets* (Wiley, Chichester, 1994)
23. S. Haykin, *Neural Networks: A Comprehensive Foundation* (Prentice-Hall, Englewood Cliffs, 1999)
24. R. Reyna, A. Giralt, D. Esteve, Head detection inside vehicles with a modified SVM for safer airbags, in *2001 IEEE Intelligent Transportation Systems, 2001. Proceedings*, Oakland (2001), pp. 268–272
25. D. Gao, J. Zhou, L. Xin, SVM-based detection of moving vehicles for automatic traffic monitoring, in *2001 IEEE Intelligent Transportation Systems, 2001. Proceedings*, Oakland (2001) pp. 745–749
26. K.-R. Müller, A. Smola, G. Rätsch, B. Schölkopf, J. Kohlmorgen, V. Vapnik, Predicting time series with support vector machines, in *Artificial Neural Networks – ICANN'97*, ser. Lecture Notes in Computer Science, vol. 1327 (Springer, Berlin/Heidelberg, 1997), pp. 999–1004
27. C. Chatfield, *The Analysis of Time Series: An Introduction* (CRC Press, Boca Raton, 2013)
28. L. Cherkasova, M. Gupta, Characterizing locality, evolution, and life span of accesses in enterprise media server workloads, in *Proceedings of the 12th International Workshop on Network and Operating Systems Support for Digital Audio and Video*, ser. NOSSDAV '02 (ACM, Miami Beach, 2002), pp. 33–42
29. C.-H. Wu, J.-M. Ho, D. Lee, Travel-time prediction with support vector regression. IEEE Trans. Intell. Trans. Syst. **5**(4), 276–281 (2004)
30. A. Sorjamaa, J. Hao, N. Reyhani, Y. Ji, A. Lendasse, Methodology for long-term prediction of time series. Neurocomputing **70**(16–18), 2861–2869 (2007)
31. A. Smola, B. Schölkopf, Statistics and computing. Tutor. Support Vector Regress. **14**(3), 199–222 (2004)
32. F. Ye, Z. Zhang, K. Chakrabarty, X. Gu, Board-level functional fault diagnosis using multikernel support vector machines and incremental learning. IEEE Trans. Comput. Aided Design Integr. Circuits Syst. **33**(2), 279–290 (2014)
33. F. Ye, Z. Zhang, K. Chakrabarty, X. Gu, Board-level functional fault diagnosis using artificial neural networks, support-vector machines, and weighted-majority voting. IEEE Trans. Comput. Aided Des. Integr. Circuits Syst. **32**(5), 723–736 (2013)
34. E. Frank, M. Mayo, S. Kramer, Alternating model trees, in *Proceedings of the 30th Annual ACM Symposium on Applied Computing*, ser. SAC '15 (ACM, New York, 2015). [Online]. Available: http://dx.doi.org/10.1145/2695664.2695848

35. M. Hall, E. Frank, G. Holmes, B. Pfahringer, P. Reutemann, I. H. Witten, The weka data mining software: an update. SIGKDD Explor. **11**(1), 10–18 (2009). Available: http://www.cs.waikato. ac.nz/ml/weka/
36. L.J. Cao, F. Tay, Support vector machine with adaptive parameters in financial time series forecasting. IEEE Trans. Neural Netw. **14**(6), 1506–1518 (2003)
37. R. Hyndman, Time series data library (2015). [Online]. Available: https://datamarket.com/ data/list/?q=provider:tsdl

Chapter 7
Conclusion

7.1 Book Contributions

This thesis has presented operation optimization and knowledge discovery solutions
for an EIS. Topics covered include production workflow optimization, operation
scheduling, resource allocation, process-execution time and process status prediction, order/service fulfillment prediction, and enterprise service-level performance
analysis and prediction. In contrast to state-of-the-art methods, which are based on
stand-alone statistical methods, analytical methods, or machine-learning algorithms,
this thesis has focused on designing unified real-time and data-driven applications
that integrate statistical methods and machine-learning algorithms. Therefore,
correlated objectives of an EIS can be optimized in a comprehensive data-driven
framework.

A number of realistic issues have been considered for data-driven operation
optimization and knowledge discovery techniques of an EIS. Due to heterogeneity
of resources, stochastic resource malfunctions, time-varying resource queue status,
and diverse process attributes, a fundamental design challenge is the estimation of
process-execution time. We have presented two different techniques to solve the
this challenge. First, a risk-aware process-execution time estimation model has been
built to incorporate heterogeneous, transient, and stochastic factors. The reliability
of this model is based on thorough understanding and highly accurate simulations
of a real enterprise production environment. Second, a machine-learning and
statistical-analysis integrated model has been built to perform the prediction. Such
data-driven approach enables more flexibility by releasing the need of establishing
parametric relationships between model features. It also overcomes the limitation
caused by making assumptions as opposed to stand-alone statistical methods.

© Springer International Publishing Switzerland 2015
Q. Duan et al., *Data-Driven Optimization and Knowledge Discovery
for an Enterprise Information System*, DOI 10.1007/978-3-319-18738-9_7

Another key design challenge is the realization of real-time and simultaneous scheduling and resource allocation for a large number of heterogeneous processes, which has been intensified by the increase in volume and diversity of demands. An evolutionary algorithm based production scheduler has been designed to address this challenge. It first acquires optimal order dispatching priority at the global level and next obtains detailed process priority and resource allocation at a lower level of granularity. The production scheduler handles large volume of orders in a distributed manner and deals with new orders with minimal impact on existing solutions. With superior performance compared to current solutions, the production scheduler has highlighted its potential benefit in reducing or eliminating production inefficiencies and enhancing the productivity of an enterprise.

The event-log databases of an EIS record major activities occurred on the service and production environment. It is of great value to the realization of different knowledge discoveries for an enterprise.

First, accurate predictions of both process-execution time and process status are crucial for the development of an intelligent EIS. We have proposed new process-execution time-prediction and process status-prediction methods for an EIS by integrating statistical methods with machine-learning algorithms. Comparison results obtained from the real-life data of an enterprise show that the proposed time-prediction method reduces both the relative mean error and the root-mean-squared error of predictions. Furthermore, the proposed status-prediction method not only achieves higher classification accuracy than both state-of-the-art methods, it also estimates the probability of the predicted status. In addition, algorithm development and training phases of the proposed methods do not rely on any arbitrary predictive horizon. Therefore, a single time-prediction model as proposed is sufficient for status prediction as opposed to a baseline status-prediction method that requires classification models for all potential predictive horizons.

Second, optimization of order-admission or service-establishment processes for an enterprise requires accurate prediction of order life cycle and fulfillment status. We have developed an intelligent order-admission framework that provides admission decisions in real-time for new orders using machine-learning and decision-integration techniques. The framework consists of three classifiers: Support Vector Machine (SVM), Decision Tree (DT), and Bayesian Probabilistic Model (BPM). The classifiers are trained by history orders and used to predict completion status for new orders. A decision integration technique is implemented to combine the results of the classifiers and predict due dates. Experimental results derived using real enterprise data show that the order completion-status prediction accuracy is significantly improved by the decision-integration strategy. The proposed multi-classifier model also outperforms a stand-alone regression model.

Finally, superior enterprise service-level performance is the collective goal of data-driven applications in an EIS. Analytical and predictive models of enterprise service-level time series within an EIS can provide meaningful insights into potential business problems and generate guidance for appropriate solutions. We have proposed a new univariate method in dealing with mid-term time-series prediction. The proposed method first analyzes the hierarchical periodic structure in one time series and decomposes it into trend, season, and noise components. By discarding the noise component, the proposed method only focuses on predicting repetitive season and smoothed trend components. As a result, this method significantly improves upon the performance of baseline methods in mid-term time-series prediction. Moreover, we have proposed a new multivariate method in dealing with short-term time-series prediction. The proposed method utilizes cross-correlation information derived from multiple time series. The amount of data taken from each time series for training the regression model is determined by results from hierarchical cross-correlation analysis. Such a data-filtering strategy leads to improved algorithm efficiency and prediction accuracy. In conclusion, by combining statistical methods with advanced machine-learning algorithms, we have achieved a significantly superior performance in both short-term and mid-term time-series predictions compared to state-of-the-art (baseline) methods.

Appendix A
Derivation of Eq. (3.3)

This appendix presents a derivation of Eq. (3.3) that was used in Chap. 3.

When $N = 0$, Eq. (3.3) equals zero. When N approaches infinity, Eq. (3.3) should also approaches infinity.

Define:

$$L = \frac{E[X_{N+1}]}{E[X_N]} = \frac{\sum_{i=1}^{N+1} \frac{w_i}{\prod_{k=i}^{N+1}(1-\gamma_k)}}{\sum_{i=1}^{N} \frac{w_i}{\prod_{k=i}^{N}(1-\gamma_k)}} \tag{A.1}$$

$$= \frac{\frac{1}{1-\gamma_{N+1}} \cdot \left[\sum_{i=1}^{N} \frac{w_i}{\prod_{k=i}^{N}(1-\gamma_k)} + \frac{w_{N+1}}{\prod_{k=i}^{N}(1-\gamma_k)} \right]}{\sum_{i=1}^{N} \frac{w_i}{\prod_{k=i}^{N}(1-\gamma_k)}} \tag{A.2}$$

$$= \frac{1}{1-\gamma_{N+1}} \left[1 + \frac{\frac{w_{N+1}}{\prod_{k=i}^{N}(1-\gamma_k)}}{\sum_{i=1}^{N} \frac{w_i}{\prod_{k=i}^{N}(1-\gamma_k)}} \right] \tag{A.3}$$

$$> \frac{1}{1-\gamma_{N+1}} \tag{A.4}$$

$$> 1. \tag{A.5}$$

Since Eq. (3.3) monotonically increases as N increases with a rate greater than 1, therefore, Eq. (3.3) diverges with N.

© Springer International Publishing Switzerland 2015
Q. Duan et al., *Data-Driven Optimization and Knowledge Discovery for an Enterprise Information System*, DOI 10.1007/978-3-319-18738-9

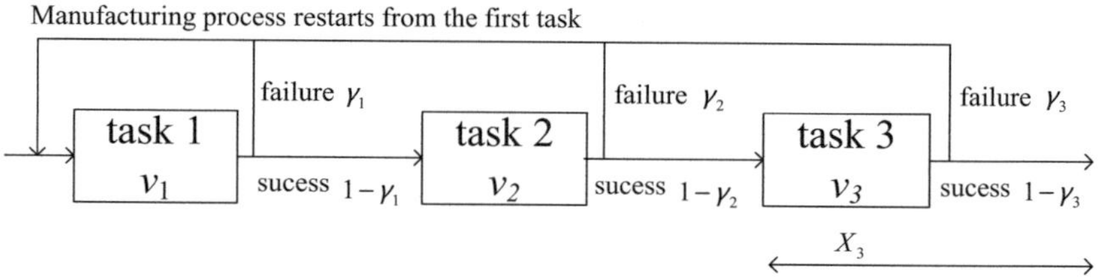

Manufacturing process restarts from the first task

Fig. A.1 Manufacturing process for a 3-stage single-part product

An illustration of Fig. 3.2 with three sequential tasks is shown in Fig. A.1. Variables X_1, X_2, and X_3 represent the time to complete all three tasks, the last two tasks, the last task, respectively. They are related by the follow equations:

$$E[X_3] = w_3 + \gamma_3 \cdot E[X_1] + (1 - \gamma_3) \cdot 0 \tag{A.6}$$

$$E[X_2] = w_2 + \gamma_2 \cdot E[X_1] + (1 - \gamma_2) \cdot E[X_3] \tag{A.7}$$

$$E[X_1] = w_1 + \gamma_1 \cdot E[X_1] + (1 - \gamma_1) \cdot E[X_2] \tag{A.8}$$

by solving Eqs. (A.6)–(A.8), we can obtain:

$$E[X_1] = \frac{w_1}{(1 - \gamma_1)(1 - \gamma_2)(1 - \gamma_3)} + \frac{w_2}{(1 - \gamma_2)(1 - \gamma_3)} + \frac{w_3}{1 - \gamma_3} \tag{A.9}$$

$$= \sum_{i=1}^{3} \frac{w_i}{\prod_{k=i}^{3} \left(1 - \gamma_k\right)}. \tag{A.10}$$

If the number of stages is extended to N, by replacing 3 by N in Eq. A.10, we can get the same expression as Eq. (3.3), which is repeated here:

$$E[X] = \sum_{i=1}^{N} \frac{w_i}{\prod_{k=i}^{N} \left(1 - \gamma_k\right)}. \tag{A.11}$$

Therefore, Eq. (3.3) holds true for all the values of N.

Equation (3.3) can be validated by mathematical induction as follows:

In the base case where $N = 1$, according to the definition of expectation, the expectation, $E[X_1]$, is defined by [1]:

$$E[X_1] = \sum_{i} x_i p_X(x_i) \tag{A.12}$$

Fig. A.2 Manufacturing process for a $(N + 1)$-stage single-part product

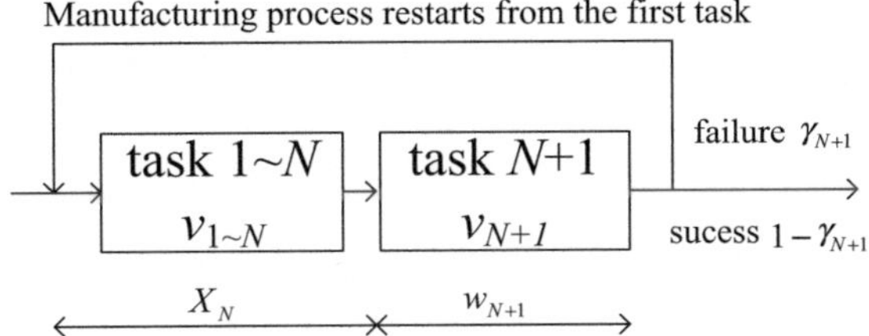

$$= \sum_{q_1=0}^{+\infty} (q_1 + 1) \cdot w_1 \cdot \gamma_1^{q_1} \cdot (1 - \gamma_1) \tag{A.13}$$

$$= w_1 \cdot (1 - \gamma_1) \sum_{q_1=0}^{+\infty} (q_1 + 1) \cdot \gamma_1^{q_1} \tag{A.14}$$

$$= \frac{w_1}{1 - \gamma_1} \tag{A.15}$$

where x_i is given by Eq. (B.3) and $p_X(x_i)$ is given by Eq. (B.7). Therefore, Eq. (3.3) is correct for $N = 1$.

Assume Eq. (3.3) to be true for some $N \geq 2$. If one more stage is added at the end of the chain, then there are $N + 1$ stages. Figure A.2 shows the case when there are $N + 1$ sequential stages. The first N stages are simplified as one stage. Variables X_N and X_{N+1} represent the time to complete the first N tasks and the time to complete all $N + 1$ tasks, respectively. They are related as follows:

$$E[X_{N+1}] = E[X_N] + w_{N+1} + (1 - \gamma_{N+1}) \cdot 0 + \gamma_{N+1} \cdot E[X_{N+1}] \tag{A.16}$$

$$= \frac{E[X_N]}{1 - \gamma_{N+1}} + \frac{w_{N+1}}{1 - \gamma_{N+1}} \tag{A.17}$$

by plugging in Eq. (3.3) into Eq. (A.17), we get:

$$E[X_{N+1}] = \sum_{i=1}^{N} \frac{w_i}{\prod_{k=i}^{N+1} \left(1 - \gamma_k\right)} + \frac{w_{N+1}}{1 - \gamma_{N+1}} \tag{A.18}$$

$$= \sum_{i=1}^{N+1} \frac{w_i}{\prod_{k=i}^{N+1} \left(1 - \gamma_k\right)}. \tag{A.19}$$

Therefore, if Eq. (3.3) is true for some $N \geq 2$, it is also true for $N + 1$. Since it has been proved to be true for $N = 1$, by mathematical induction, Eq. (3.3) holds true for all the values of N.

Appendix B
Derivation of the PMF of Random Variable X

In this appendix, we present a derivation of the probability mass function (PMF) of random variable X described in Sect. 3.2.2.

Random variable Q_i is the number of failures at task v_i. The value of Q_i is denoted by q_i. Failures are independent events. The joint probability of the event that $Q_1 = q_1, \ldots,$ and $Q_N = q_N$ is computed by Eq. (B.1):

$$P(Q_1{=}q_1,\ldots,Q_N{=}q_N) = \frac{\left(\sum_{i=1}^{N} q_i\right)!}{\prod_{i=1}^{N} q_i!} \cdot \prod_{i=1}^{N} \gamma_i^{q_i} \cdot \prod_{i=1}^{N}(1-\gamma_i)^{(\sum_{k=i+1}^{N} q_k+1)} \qquad \text{(B.1)}$$

where γ_i is the failure probability of task v_i (defined in Eq. (3.2)).

Discrete random variable X is the execution time of a chain of sequential tasks. The value of X, which is denoted by x, depends on the values of Q_1 to Q_N as follows:

$$X = \sum_{i=1}^{N} \left[\left(\sum_{k=i}^{N} Q_k + 1 \right) \times w_i \right] \qquad \text{(B.2)}$$

where w_i is the execution time of task v_i (defined in Eq. (3.1)).

Therefore, a *combination* of values $(q_1, \ldots, q_N)$ for variables $(Q_1, \ldots, Q_N)$ can result in a value x for variable X. Note that some combinations of $(q_1, \ldots, q_N)$ can result in the same value of X. Set $\Phi_{X=x}$ contains all combinations such that they all result in the same x for X. Therefore, value x is computed as follows:

$$x = \sum_{i=1}^{N} \left[\left(\sum_{k=i}^{N} q_k + 1 \right) \times w_i \right], \quad (q_1, \ldots, q_N) \in \Phi_{X=x}. \qquad \text{(B.3)}$$

© Springer International Publishing Switzerland 2015

Q. Duan et al., *Data-Driven Optimization and Knowledge Discovery for an Enterprise Information System*, DOI 10.1007/978-3-319-18738-9

The probability of the event that $X = x$ equals the sum of the probabilities of events $Q_1 = q_1, \ldots, Q_N = q_N$, for all $(q_1, \ldots, q_N) \in \Phi_{X=x}$. The PMF of X can be expressed as follows:

$$p_X(x) = P(X = x) \tag{B.4}$$

$$= \sum_{(q_1,\ldots,q_N)\in\Phi_{X=x}} P(Q_1 = q_1, \ldots, Q_N = q_N). \tag{B.5}$$

When $N \geq 2$, by plugging in Eq. (B.1), Eq. (B.5) can be written as:

$$p_X(x) = \sum_{(q_1,\ldots,q_N)\in\Phi_{X=x}} \left[\frac{\left(\sum_{i=1}^{N} q_i\right)!}{\prod_{i=1}^{N} q_i!} \cdot \prod_{i=1}^{N} \gamma_i^{q_i} \cdot \prod_{i=1}^{N}(1 - \gamma_i)^{(\sum_{k=i+1}^{N} q_k+1)} \right] \tag{B.6}$$

whereas when $N = 1$, Eq. (B.5) can be written as:

$$p_X(x) = \gamma_1^{q_1} \cdot (1 - \gamma_1). \tag{B.7}$$

When $N = 1$ (i.e., only one task is in the product), value $x = (q_1 + 1) \cdot w_1$, which means x equals multiples of w_1. No matter how many times task v_1 fails, as soon as it succeeds, the one-task product gets manufactured. Term $\gamma_1^{q_1} \cdot (1 - \gamma_1)$ in Eq. (B.7) is the joint probability of the q_1 times failures and the one time success.

When $N \geq 2$, value x is a linear combination of w_1 to w_N as shown in Eq. (B.3). For every combination $(q_1, \ldots, q_N) \in \Phi_{X=x}$, although they derive the same value x, they represent different situations of failures. Therefore, a combination $(q_1, \ldots, q_N)$ represents a *failure situation*. For each failure situation, set Δ contains all failure sequences under this situation. A *failure sequence* describes the sequence of failures, i.e., the sequence in which failures occur. In an example when $N = 2$, there are two sequential tasks v_1 and v_2 that must be executed to manufacture a product. One possible failure situation is $(q_1 = 2, q_2 = 1)$. It means that task v_1 fails twice and task v_2 fails once. Set Δ under this situation contains three failure sequences: $v_1 - v_1 - v_2$, $v_1 - v_2 - v_1$, and $v_2 - v_1 - v_1$. Therefore, $|\Delta| = 3$. They all result in the same value x by Eq. (B.3):

$$x = (2 + 1 + 1) \cdot w_1 + (1 + 1) \cdot w_2$$

even though they are different failure sequences.

Failures at the same task have no difference. Term $(\sum_{i=1}^{N} q_i)!$ in Eq. (B.6) is the number of failure sequences only if all failures are different. Term $(\sum_{i=1}^{N} q_i)!/\prod_{i=1}^{N} q_i!$ is the actual number of failure sequences. It is derived by excluding indistinguishable failure sequences caused by multiple failures at the same task from term $(\sum_{i=1}^{N} q_i)!$. In the $N = 2$ example, for failure situation $(q_1 = 2, q_2 = 1)$, the number of distinct failure sequences is obtained by:

$$\begin{aligned}
|\Delta| &= \frac{(q_1 + q_2)!}{q_1! q_2!} \\
&= \frac{(2 + 1)!}{2! 1!} \\
&= 3.
\end{aligned}$$

According to the rework policy, every failure at task v_i, where $1 \leq j < i \leq N$, means a success at task v_j. Term $\sum_{k=i+1}^{N} q_k$ is the number of successes at task v_i caused by failures at task v_k, where $i < k \leq N$. Note that, a product can only be manufactured when all tasks uninterruptedly succeed from the first to the last task. Therefore, task v_i must succeed at least once. Term $(\sum_{k=i+1}^{N} q_k + 1)$ in Eq. (B.6) is the total number of successes at task v_i. Hence term $\prod_{i=1}^{N}(1 - \gamma_i)^{\sum_{k=i+1}^{N} q_k + 1}$ is the joint probability of successes at all tasks. Term $\prod_{i=1}^{N} \gamma_i^{q_i}$ in Eq. (B.6) is the joint probability of failures at all tasks.

Due to the independence of failures, the probability of any failure sequence in the set Δ under the same failure situation is the same. A failure sequence implies successes, therefore, term $\prod_{i=1}^{N} \gamma_i^{q_i} \cdot \prod_{i=1}^{N}(1 - \gamma_i)^{\sum_{k=i+1}^{N} q_k + 1}$ is the probability of a failure sequence in set Δ under failure situation $(q_1, \ldots, q_N)$.

In summary, when $N \geq 2$, term $(\sum_{i=1}^{N} q_i)! / \prod_{i=1}^{N} q_i!$ is the number of failure sequences under failure situation $(q_1, \ldots, q_N)$; term $\prod_{i=1}^{N} \gamma_i^{q_i} \cdot \prod_{i=1}^{N}(1 - \gamma_i)^{\sum_{k=i+1}^{N} q_k + 1}$ is the probability of a failure sequence under failure situation $(q_1, \ldots, q_N)$; and set $\Phi_{X=x}$ contains all failure situations that result in $X = x$. Therefore, the probability of $X = x$ equals the sum of the probabilities of failure sequences under each failure situation in set $\Phi_{X=x}$. It can be calculated by Eq. (B.6). Hence Eq. (B.5) is the PMF of random variable X.

Appendix C
Derivation of Eq. (3.4)

In this appendix, we present a derivation of the expectation of random variable W that is described in Sect. 3.2.2.

C.1 Approximate the Distribution of X by an Exponential Distribution

Discrete random variable X is the execution time of a chain of sequential tasks. As discussed in Sect. B, Eq. (B.5) is the PMF of random variable X. Appendix A discussed that the expectation of X can be computed by Eq. (3.3). If task execution time w_i at each stage i is small enough, then the intervals between possible values of X are also small. Therefore, the execution time X can be viewed as a continuous random variable. We observed that, the probability mass function of X can be approximated by the probability density function of an exponentially distributed random variable. The rate of the exponential distribution equals $1/E[X]$. Figures C.1 and C.2 demonstrate such approximation.

C.2 The Expectation of the Maximum of Exponentials

Let random variables $W_1, \ldots, W_H$ be independent exponentially distributed random variables with the same rate $\lambda = 1/E[X]$, $Y = \max(W_1, \ldots, W_H)$. It is proved in [2] that the expectation of Y is given by the follow equation:

$$E[Y] = \frac{1}{\lambda} \cdot \left(1 + \frac{1}{2} + \ldots + \frac{1}{H} \right) \tag{C.1}$$

© Springer International Publishing Switzerland 2015
Q. Duan et al., *Data-Driven Optimization and Knowledge Discovery for an Enterprise Information System*, DOI 10.1007/978-3-319-18738-9

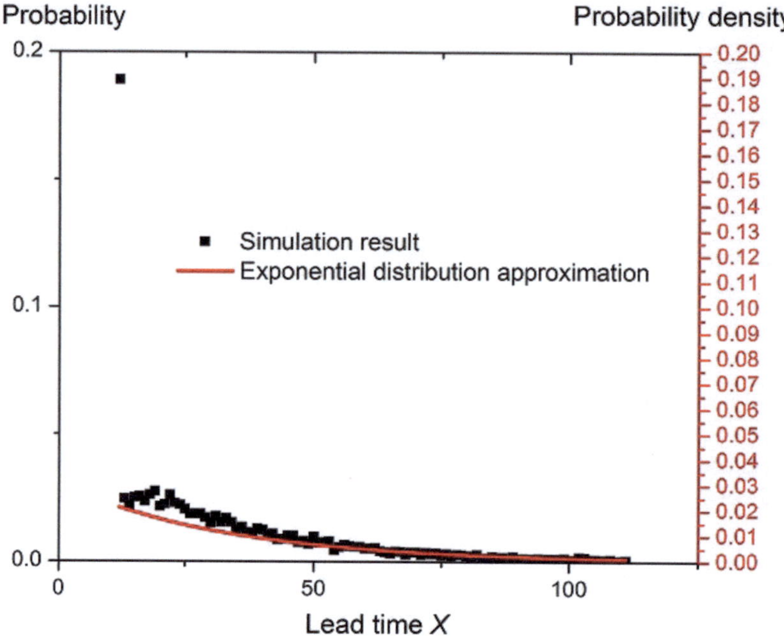

Fig. C.1 The probability mass function derived from simulation results (*left*) and the probability density function approximated by an exponential distribution (*right*) for a 12-stage chain of sequential tasks with task execution times less than 1 time unit

$$= E[X] \cdot \sum_{j=1}^{H} \frac{1}{j}. \tag{C.2}$$

Based on the discussed in Sect. C.1 and a worst case estimation, we approximate the distributions of execution times for parallel parts with an identical exponential distribution whose rate equals $\lambda = \max(1/E[W_j])$, where $j \in \{1, \ldots, H\}$. Hence we estimate the execution time of all the parts by Eq. (3.4), which is repeated here:

$$E[Y] \approx \max_{j \in \{1,\ldots,H\}} \{E[W_j]\} \cdot \sum_{j=1}^{H} \frac{1}{j}. \tag{C.3}$$

We can obtain $E[Y]$ from Y's PMF and CDF, which requires the PMF and CDF of each W_j. Rigorous computation is too complicated to be realized in real time. Therefore, we adopt Eq. (3.4) for simplicity and fast computation.

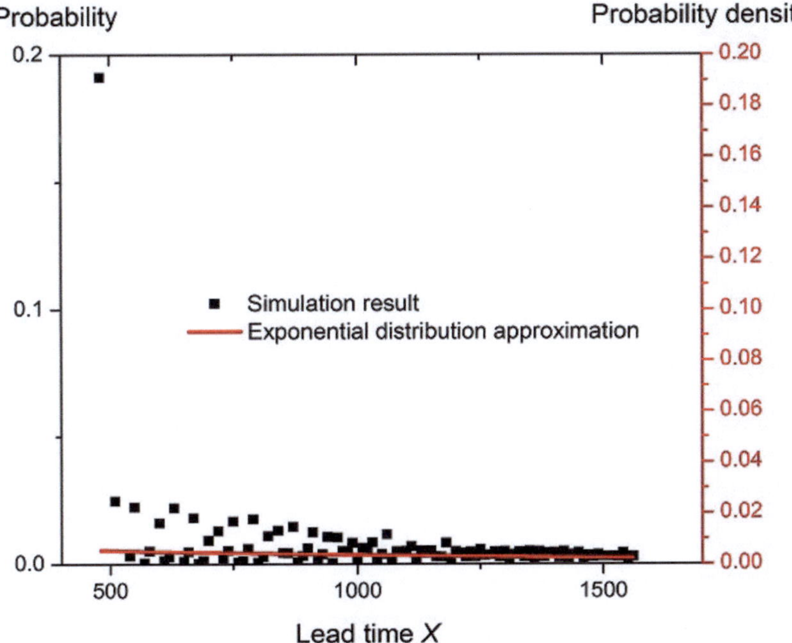

Fig. C.2 The probability mass function derived from simulation results (*left*) and the probability density function approximated by an exponential distribution (*right*) for a 12-stage chain of sequential tasks with task execution times no less than 30 time units

Appendix D
Introduction to SVR

Support-vector machines (SVMs) [3] have been extensively used for classification and regression [4, 5]. The key idea of an SVM is based on computing a linear regression function in a high-dimensional feature space by mapping the input data via a kernel function. It has been shown that an SVM offers major advantages in high-dimensionality space since its optimization is less dependent on the dimensionality of the input space. Compared to the empirical risk minimization (ERM) principle adopted in neural networks, the structural risk minimization (SRM) principle in SVM has greater generation ability [6]. It is robust in recognizing subtle patterns from complex data sets, even with corrupted data. Applications in pattern recognition [7, 8], vision-based head recognition [9] and vehicle detection [10] have already demonstrated that SVM is rigorous from a theoretical point of view and can lead to superior performance and great potential.

The application of SVM to regression is called support vector regression (SVR). The tutorial in [11] describes the process of selecting algorithms for training SVR models. We are therefore motivated by the many successful applications of SVR models (discussed in Sect. 4.1). We have found that SVR achieves the highest prediction accuracy for our testing data as compared with other algorithms such as linear regression and neural networks.

We first provide a brief introduction to support vector machine (SVM) in regression approximation [12, 13]. The basic idea of SVM is to map the training data from the input space into a high-dimensional feature space via a function ϕ and then construct a separating hyperplane with maximum margin in the feature space. The set of training data points is denoted as (x_1, y_1), (x_2, y_2), ..., (x_l, y_l), where $x_i \in X \subseteq R^n$, $y_i = \pm 1$ denotes two class labels, and l is the number of training data. An SVM will find a hyperplane direction vector w and an offset scalar b such that $f(x) = w \cdot \phi(x) + b \geq 0$ for positive samples and $f(x) = w \cdot \phi(x) + b \leq 0$ for negative samples, which is illustrated in Fig. D.1. Therefore, even if a linear function cannot

© Springer International Publishing Switzerland 2015
Q. Duan et al., *Data-Driven Optimization and Knowledge Discovery
for an Enterprise Information System*, DOI 10.1007/978-3-319-18738-9

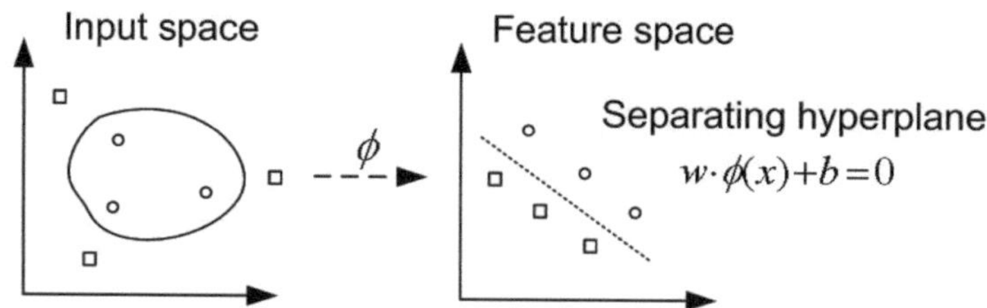

Fig. D.1 The basic idea of SVM is to find a hyperplane in high-dimensional feature space that can separate instances into two classes

be found in the input space to decide which class the given data belongs to, the function of ϕ can be used to find an optimal hyperplane that can clearly distinguish between instances of two classes.

SVR is derived from the idea of an SVM. The set of training data points is denoted as (x_1, y_1), (x_2, y_2), ..., (x_l, y_l), where $x_i \in X \subseteq R^n$, $y_i \in Y \subseteq R$, l is the number of training data. These data points are randomly and independently generated from the same distributions. An SVM approximates the decision function using the following form:

$$f(x) = w \cdot \phi(x) + b \tag{D.1}$$

where $\phi(x)$ represents the high-dimensional feature space that is obtained through a nonlinear mapping from the input space x. The hyperplane direction vector w and offset scalar b are estimated by minimizing the regularized risk function (D.2):

$$\text{minimize} \quad R_{reg}(f) = \frac{1}{2} \|w\|^2 + C \frac{1}{l} \sum_{i=1}^{l} L_\varepsilon(y_i, f(x_i)) \tag{D.2}$$

$$L_\varepsilon(y_i, f(x_i)) = \begin{cases} |y_i - f(x_i)| - \varepsilon, & |y_i - f(x_i)| \geq \varepsilon \\ 0, & \text{otherwise.} \end{cases} \tag{D.3}$$

The first term $\|w\|^2$ of the regularized risk function $R_{reg}(f)$ is called the regularized term. Minimizing $\|w\|^2$ makes the risk function as flat as possible, thus playing the role of controlling the function regularization capacity. Function $L_\varepsilon(\cdot)$ is the loss function [14]. The second term $(1/l) \sum_{i=1}^{l} L_\varepsilon(y_i, f(x_i))$ is the empirical error measured by the ε-insensitive loss function. The loss function provides the advantage of using sparse data points to represent the regularized risk function (D.2). Parameter C is the regularization factor, which determines the penalty of a data point at the wrong side of the hyperplane, and ε is called the tube size, which determines the data inside the ε tube to be ignored in regression. They are user-prescribed parameters and need to be empirically optimized while building the model.

In order to get estimates of w and b, the regularized risk function (D.2) is transformed to the primal objective function (D.4) by introducing the positive slack variables ξ_i and ξ_i^*, which are used to measure errors outside of the ε tube.

$$\text{minimize} \quad R_{reg}(f) = \frac{1}{2}\|w\|^2 + C\sum_{i=1}^{l}\left(\xi_i + \xi_i^*\right)$$

$$\text{subject to} \quad y_i - w \cdot \phi(x_i) - b \leq \varepsilon + \xi_i$$

$$w \cdot \phi(x_i) + b - y_i \leq \varepsilon + \xi_i^*, \qquad i = 1, \cdots, l$$

$$\xi_i^{(*)} \geq 0 \tag{D.4}$$

where $\xi_i^{(*)}$ denotes variables with and without $*$.

Finally, by introducing Lagrange multipliers and exploiting the optimality constraints, the decision function (D.1) has the following explicit form:

$$f(x) = \sum_{i=1}^{l}\left(a_i - a_i^*\right)K(x_i, x) + b \tag{D.5}$$

In function (D.5), $a_i^{(*)}$ are Lagrange multipliers. They satisfy the equalities $a_i \times a_i^* = 0$, and $a_i^{(*)} \geq 0$ where $i = 1, \cdots, l$, and they are obtained by maximizing the dual of the primal objective function (D.4), which has the following form:

$$W\left(a_i^{(*)}\right) = \sum_{i=1}^{l} y_i\left(a_i - a_i^*\right) - \varepsilon \sum_{i=1}^{l} y_i\left(a_i + a_i^*\right)$$

$$-\frac{1}{2}\sum_{i=1}^{l}\sum_{j=1}^{l}\left(a_i - a_i^*\right)\left(a_j - a_j^*\right)K(x_i, x_j) \tag{D.6}$$

with the following constraints:

$$\sum_{i=1}^{l}\left(a_i - a_i^*\right) = 0, \qquad 0 \leq a_i^{(*)} \leq C, \ i = 1, \cdots, l.$$

Vector w is written in terms of data points as:

$$w = \sum_{i=1}^{l}\left(a_i - a_i^*\right)\phi(x_i) \tag{D.7}$$

Table D.1 Common kernel functions [15]

Kernel	Function $K(x_i, x_j)$	Parameter
Linear	$x_i \cdot x_j$	
Polynomial	$(x_i \cdot x_j + 1)^d$	d
Gaussian	$\exp(-(1/\sigma^2)(x_i - x_j)^2)$	σ
Radial Basis Function (RBF)	$\exp(-\gamma(x_i - x_j)^2)$	γ

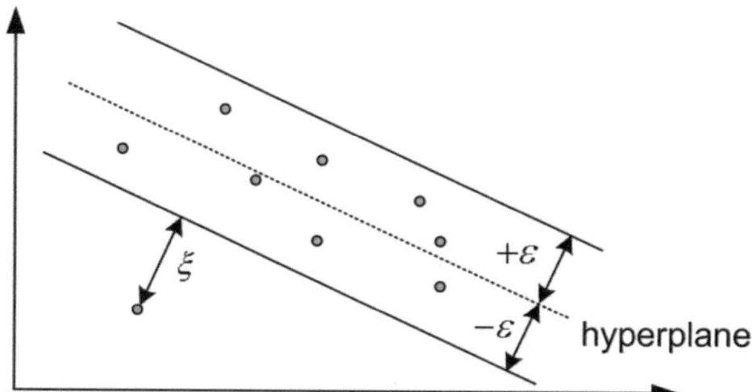

Fig. D.2 SVR to fit a tube with radius ε to the data and positive slack variables ξ measuring the data points lying outside of the tube

Function $K(x_i, x_j)$ is defined as the kernel function [15]. The value of the kernel is equal to the inner product of two vectors x_i and x_j in the feature space $\phi(x_i)$ and $\phi(x_j)$, that is, $K(x_i, x_j) = \phi(x_i) \cdot \phi(x_j)$. The reason for applying kernel function is that SVR can perform a dot product in high-dimensional feature space using low-dimensional space data without knowing the transformation ϕ. Any function that satisfies Mercer's condition [16] can be used as the kernel function.

Some commonly used kernel functions are shown in Table D.1. We experimented with these kernels and the RBF kernel demonstrated the best performance.

After solving the value of w in terms of the Lagrange multipliers, variable b can be computed by applying the Karush-Kuhn-Tucker (KKT) conditions [17]. Based on the KKT conditions, only a few coefficients $(a_i - a_i^*)$ in decision function (D.5) have nonzero values, and the corresponding training data points have approximation errors equal to or larger than the tube with radius ε, and these data points are referred as support vectors. For all points inside the ε tube, the Lagrange multipliers that equal to zero do not contribute to the decision function. Lagrange multipliers may be nonzero values and used as support vectors only if the requirement $\varepsilon_i \leq |f(x_i) - y_i| \leq \xi_i$ is satisfied (shown in Fig. D.2). In this case, the product of the Lagrange multipliers and constrains has to be equal to 0.

$$a_i \left(\varepsilon + \xi_i - y_i + (w, x_i) + b\right) = 0$$

$$a_i^* \left(\varepsilon + \xi_i^* + y_i - (w, x_i) - b\right) = 0$$

$$(C - a_i^{(*)})\xi_i^{(*)} = 0 \qquad\qquad (D.8)$$

Since $a_i^{(*)} = 0$, and $\xi_i^{(*)} = 0$ for $a_i^{(*)} \in (0, C)$, b can be computed as:

$$b = y_i - (w, x_i) - \xi_i \quad \text{for } a_i \in (0, C)$$
$$b = y_i - (w, x_i) + \xi_i \quad \text{for } a_i^* \in (0, C). \tag{D.9}$$

Therefore, as seen from the decision function (D.5), only the support vectors are used to determine the function value, as the values of $\left(a_i - a_i^*\right)$ for the other training data points are equal to zero. This highlights the main advantage of SVR, i.e., support vectors are usually only a small subset of the training data points; this is referred to as the sparsity of the solution.

In summary, in order to train a SVR model, we need to optimize the kernel function $K(\cdot)$, the regularization factor C, and the tube size ε through the procedures described in tutorial such as [11].

References

1. K. Trivedi, *Probability and Statistics with Reliability, Queuing and Computer Science Application*, 2nd edn. (Wiley, New York, 2001)
2. [Online].   Available: http://www.stat.berkeley.edu/~mlugo/stat134-f11/exponential-maximum.pdf
3. H. Drucker, J. Burges, L. Kaufman, A. Smola, V. Vapnik, *Advances in Neural Information Processing Systems*, 9th edn. (MIT Press, Cambridge, 1997)
4. S.R. Gunn, Support vector machine for classification and regression. Technical report, University of Southampton, Southampton (1998)
5. K.-R. Müller, A.J. Smola, G. Rätsch, B. Schölkopf, J. Kohlmorgen, V. Vapnik, Predicting time series with support vector machines, in *International Conference on Artificial Neural Networks*, Huston. Springer Lecture Notes in Computer Science (1997), pp. 999–1004
6. U. Thissen, R. van Brakel, A. de Weijer, W. Melssen, L. Buydens, Using support vector machines for time series prediction. Chemom. Intell. Lab. Syst. **69**(1–2), 35–49 (2003)
7. V. Vapnik, An overview of statistical learning theory. IEEE Trans. Neural Netw. **10**(5), 988–999 (1999)
8. J. tao Ren, X. ling Ou, Y. Zhang, D.-C. Hu, Research on network-level traffic pattern recognition, in *The IEEE 5th International Conference on Intelligent Transportation Systems, 2002. Proceedings*, Singapore (2002), pp. 500–504
9. R. Reyna, A. Giralt, D. Esteve, Head detection inside vehicles with a modified svm for safer airbags, in *2001 IEEE Intelligent Transportation Systems. Proceedings*, Oakland (2001), pp. 268–272
10. D. Gao, J. Zhou, L. Xin, SVM-based detection of moving vehicles for automatic traffic monitoring, in *2001 IEEE Intelligent Transportation Systems. Proceedings*, Oakland (2001), pp. 745–749
11. A. Smola, B. Schölkopf, Statistics and computing. Tutor. Support Vector Regress. **14**(3), 199–222 (2004)
12. C.-H. Wu, J.-M. Ho, D. Lee, Travel-time prediction with support vector regression. IEEE Trans. Intell. Trans. Syst. **5**(4), 276–281 (2004)
13. L.J. Cao, F. Tay, Support vector machine with adaptive parameters in financial time series forecasting. IEEE Trans. Neural Netw. **14**(6), 1506–1518 (2003)

14. N. de Freitas, M. Milo, P. Clarkson, M. Niranjan, A. Gee, Sequential support vector machines, in *Neural Networks for Signal Processing IX, 1999. Proceedings of the 1999 IEEE Signal Processing Society Workshop*, Taipei (1999), pp. 31–40
15. H.T. Lin, C.J. Lin, A study on sigmoid kernels for SVM and the training of non-PSD kernels by SMO-type methods. Technical report, Department of Computer Science and Information Engineering, National Taiwan University, Taipei (2003). http://www.csie.ntu.edu.tw/cjlin/papers/tanh.pdf
16. V.N. Vapnik, *The Nature of Statistical Learning Theory* (Springer, New York, 1995), pp. 40–422
17. H.W. Kuhn, A.W. Tucker, Nonlinear programming, in *Procedings of the 2th Berkeley Symposium on Mathematical Statistics and Probabilistics* (University of California Press, Berkeley, 1951), pp. 481–492

®
FSC
www.fsc.org